Praise for ARCTIC

I'd have died of pneumonia one winter *2
Huslia to the Tanana Hospital in rotten* ....
who . . . improved life for everyone in the ..... *Country.*
— Sidney Huntington, Galena, Alaska

I've never read a book that so skillfully puts the reader into a bush pilot's head and so compellingly captures the drama, adventure, and misadventure of Arctic flying.
— Cliff Cernick, Alaska editor, Flyer magazine

To those of us who flew the backcountry then, Andy Anderson was the bush pilot's pilot. More than just being a skilled aviator, he was a compassionate man who put the lives and cares of his passengers before his own.
— Ray Tremblay, pilot

This is one of the best accounts of a bush pilot's life and times in Alaska . . . an excellent read for everyone.
— Ed Klopp, CAA station manager, Bettles, Alaska, 1949-51

He was a non-nonsense pilot who earned the respect of his peers and that of many Indian and Eskimo villagers ... His book brings alive the beauty of the Koyukuk and its people, and his contributions to safe, dependable air service there.
— Paul Haggland, pilot, former manager,
Fairbanks International Airport

I have never read a better account of the country and the people I know so well. It is a great story of a bootstrap-up, pioneering bush pilot. Every reader will become absorbed with the story of Andy Anderson who made such a difference to the people, and in the transportation, in Alaska's Koyukuk Country.
— Prof. Ernest N. Wolff, University of Alaska Fairbanks (retired)

Andy Anderson stood out as the most generous of pilots and the most gracious of hosts.
— Tom Wardleigh, Alaska Aviation Safety Foundation

Other books by Jim Rearden include:

Tales of Alaska's Big Bears

Trail of the Eagle, with Bud Conkle

Castner's Cutthroats

Wind on the Water, with Lenora Huntley Conkle

In the Shadow of Eagles, with Rudy Billberg

White Squaw, with Delores Cline Brown

Shadows on the Koyukuk, with Sidney Huntington

Koga's Zero

Alaska's Wolf Man

ARCTIC BUSH PILOT

FROM NAVY COMBAT TO FLYING ALASKA'S NORTHERN WILDERNESS

A MEMOIR:

James "Andy" Anderson
as told to Jim Rearden

EPICENTER PRESS

Epicenter Press Inc. is a regional press founded in Alaska whose interests include but are not limited to the arts, history, environment, and diverse cultures and lifestyles of the North Pacific and high latitudes. We seek both the traditional and innovative in publishing nonfiction tradebooks and giftbooks featuring contemporary art and photography.

Text © 2000 Jim Rearden
Photos © 2000 Jim Rearden unless otherwise credited

Publisher: Kent Sturgis
Editor: Don Graydon
Mapmaker: Russell Nelson
Proofreader: Sherrill Carlson
Cover and text design, typesetting: Elizabeth Watson

ISBN 0-945397-83-6

To order single copies of ARCTIC BUSH PILOT, mail $16.95 (WA residents add $1.46 sales tax) plus $5 for Priority Mail shipping to: Epicenter Press, Box 82368, Kenmore, WA 98028; phone our 24-hour order line, 800-950-6663; or visit our website, EpicenterPress.com.

Booksellers: Retail discounts are available from our distributor, Graphic Arts Center Publishing, Box 10306, Portland, OR 97210. Phone 800-452-3032.

Printed in Canada

First printing May 2000

10 9 8 7 6 5 4

ACKNOWLEDGMENTS

My thanks to every Indian and Eskimo villager, every miner and trapper, and every bushrat who lived in the Koyukuk Valley and Brooks Range where I flew, for in one way or another each contributed to the development of scheduled air service there.

Frank Tobuk and Warren "Canuck" Killen, Wien Airlines employees at Bettles, kept me flying; Canuck was my right arm. Neither of these fine men ever failed me.

The Wien brothers, Noel, Sig, and Fritz, and the next generation of Wiens—Richard, Merrill, and Bob—all contributed wholeheartedly to my bush operation at Bettles. The Wiens taught me much about the fine art of Alaska flying.

Russell "Mac" McConnell, although he was a Bettles FAA maintenance employee, contributed much to my flying program; Mac could do anything.

My heartfelt appreciation goes to my first wife, Hannah, who did a superb job of overseeing the Bettles roadhouse operation and raising our children. My thanks to our children, Mary, David, and Phil, for their acceptance of a difficult situation.

To my wonderful parents, my thanks for a responsible Christian upbringing. To my brothers Carl and Louis and my sister Hope, my thanks for a lifetime of moral support. And to Betty, my wife of three decades, at my side in sickness and health, in good times and bad, goes my deepest appreciation. My thanks to her three children, Cheryl, Julie, and Jay, for their understanding and steadfast faith in both of us.

Thanks too to Cliff Cernick, of Anchorage, for his professional on-the-mark line editing. William O. Seymour, professor of journalism at the University of West Virginia, made helpful comments on the manuscript. Thanks to Ernest N. Wolff for permission to use biographical details on men of the Chandalar that appeared in his booklet *Frank Yasuda and the Chandalar*. Last, my appreciation to Jim Rearden for producing a lucid, accurate record of my aviation career.

—JLA

PREFACE

Half a century ago I participated in a tiny part of the aviation history of Alaska. When I established scheduled flights in the Koyukuk River valley, which straddles the Arctic Circle, I wasn't thinking of history or of being a pioneer. I simply filled a need and took advantage of an economic opportunity. But eventually the satisfaction of serving became as important to me as the economics.

I was fortunate to arrive at Bettles, where I was based, at a period when there was a great need for aerial transportation. At first, my small, inefficient planes could only haul freight of limited size. As my business grew, I flew increasingly larger and more dependable freight-carrying airplanes. When I ended my seventeen-year bush flying career, my planes were delivering some of the bigger stuff: refrigerators, 4-by-8 sheets of plywood, entire dog teams.

At first, most villages I served had no airport; when I left, most of these villages had an airport that could handle twin-engine planes. By then these villages also had good radio communication.

At first, villagers rushed to meet my plane whenever I landed, much as pioneers of the early west gathered at a depot to greet the train. By the time I left, only those who had business at the airport met my flights; airplanes had become commonplace.

I enjoyed being a bush pilot. I enjoyed flying. I enjoyed the magnificent wilderness land over which I flew. I enjoyed the people I served. For these reasons my years at Bettles seemed almost like a paid vacation.

Remembering these pleasures, and reliving my experiences by writing about them, has given me much satisfaction. I would like to believe that this review of my time as an arctic bush pilot will contribute in a small way to the history of Alaskan aviation.

James L. "Andy" Anderson

Dr. Edward Wiegand of Sandusky, Ohio, fishes in Agiakhake

above the timberline in The Brooks Range. JIM REARDEN PHOTO

CONTENTS

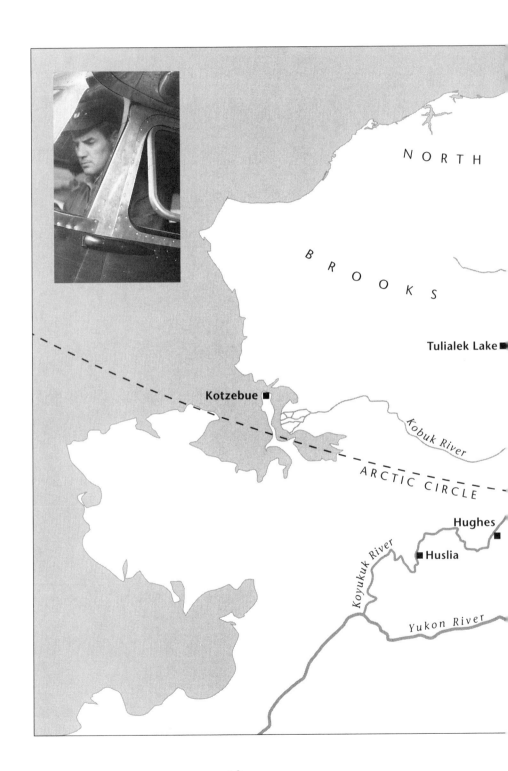

NORTH

BROOKS

Tulialek Lake■

Kotzebue■

Kobuk River

ARCTIC CIRCLE

Hughes
■

■Huslia

Koyukuk River

Yukon River

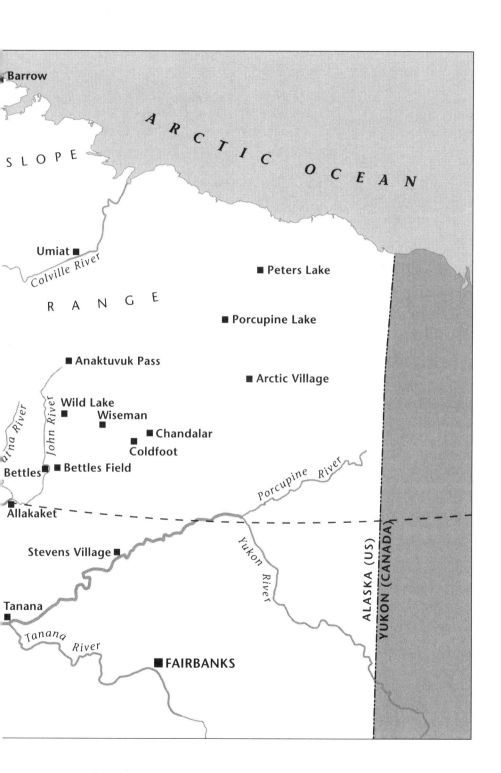

Barrow

ARCTIC

SLOPE

OCEAN

Umiat ■

Colville River

R A N G E

■ Peters Lake

■ Porcupine Lake

■ Anaktuvuk Pass

■ Arctic Village

Wild Lake
■
Wiseman
■

■ Chandalar

Coldfoot
■

Bettles ■ ■ Bettles Field

John River

na River

Porcupine River

Allakaket

Stevens Village ■

Yukon River

ALASKA (US)

YUKON (CANADA)

Tanana
■

Tanana River

■ FAIRBANKS

INTRODUCTION

BY JIM REARDEN

We were camped near a lake among the lonely cloud-shrouded peaks of the high-country, arctic Brooks Range. My partner was Dr. Neil Hosley, dean of the University of Alaska Fairbanks, and overnight we had endured high winds and an icy, driving rain. Our small tent had a few drips, but we remained dry and warm. At daylight, windblown clouds obscured nearby peaks.

At midmorning, over the sound of wind and heavy rain, came the thin drone of a low-flying small plane. We were far from any regular route. Doc and I stared at each other in disbelief. Why was an airplane flying here in such terrible weather?

As we watched, a swift, ground-hugging silver-and-green float-plane emerged from the murk. It whistled overhead in a steep bank, then feathered onto choppy Agiak Lake. I waded into the water, caught a pontoon, and held the rocking plane off the jagged beach. The pilot—slim, medium-size, wearing a baseball cap—poked a prematurely graying head out a window. "Are you fellows all right?"

Guide and writer Jim Rearden stands with a bull moose he took along the John River on the south slope of the Brooks Range in 1955. The antlers had a spread of 65 inches. I flew Rearden, who was on assignment for a national outdoor magazine, to a variety of mountain campsites.

Arriving at a lonely wilderness camp amidst a storm to ask that question was typical of James L. "Andy" Anderson. He was the Wien Airlines bush pilot who flew from the lonely upper Koyukuk River arctic outpost of Bettles. When any of "his people" were alone in the wilderness, Andy made them his concern. He had flown ninety miles into the storm-wracked mountains solely to make sure Doc Hosley and I—"his people"—were safe.

The year was 1955. Andy was thirty-three and had been flying the thinly populated Koyukuk country and Brooks Range for seven years. Few Americans had seen these almost-legendary mountains. This northern range, which stretches across arctic Alaska into Canada, was barely known even to veteran bush pilots. Maps of the Colorado-size region were sketchy.

But postwar Alaska was changing with increasing population and economic development. New long-range airplanes were becoming available. The wild and scenic Brooks Range, rich in minerals and in fish and game, beckoned to adventurous sportsmen, wilderness lovers, and miners.

I was there on assignment from *Outdoor Life* magazine to report on the range's potential as a hunting area. I had first met Andy three years earlier, during a wildlife survey with graduate wildlife student Burt Libby.

Burt and I had paddled a canoe to the Koyukuk River village of Huslia. At the time I was head of the Department of Wildlife Management at the University of Alaska Fairbanks. Upon our arrival, the local trader handed us a radio message from the pilot scheduled to pick us up: "Return to Fairbanks with Wien. The Stinson is down for repairs."

A day or so later, Andy flew a Wien Airlines Cessna 170 on floats into Huslia on his regular mail run. I don't remember much about my first meeting with Andy, except that he was cheerful and immediately made us feel comfortable in his small plane. Burt and I flew with him to Bettles, where we had a day-and-a-half wait for the Wien DC-3 mainliner to Fairbanks.

"Why don't you fellows take my riverboat upstream and see if you can find a moose?" Andy suggested.

We were strangers. The offer was unexpected and generous. We ran his outboard-powered boat upstream, shot a bull moose, and returned with the meat to take the Wien mainliner to Fairbanks.

That set the stage for my repeated returns to the Brooks Range via Bettles, with Andy's flying and the hospitality of the Bettles roadhouse he and his gracious wife Hannah operated.

In 1954 I resigned my professorship and launched a freelance outdoor writing career. I also became a registered big-game guide and it was only natural that I take clients into the Brooks Range, where Andy flew us into game-rich areas. I couldn't have asked for a better pilot or better service. He was generous, loaning equipment without hesitation, often making flights without charge.

For one hunt I borrowed his 25-foot riverboat and took two clients into the mountains far up the John River. At intervals Andy flew us fresh produce, bread, and mail, and hauled the meat of our moose and sheep to Bettles to store it in his meathouse.

After Doc Hosley returned to Fairbanks from our 1955 exploratory hunt, Andy invited me to stay with him for a week. I flew with him almost daily and learned a little about flying bush in Alaska's Arctic. His hours would have shocked a conventional eight-to-five clock-puncher.

"He sure has a lot of folks depending on him," I was told that week by Joe Blundell of Big Lake, a miner and one of the folks dependent on Andy.

Fairbanks big-game guide Warren "Tillie" Tilman, who was the second airplane mechanic to be licensed in the Territory of Alaska in the early 1930s, said, "Andy's a real pilot, and he's the kind who'll help you unload the plane, then help pack your stuff half a mile to a camp."

When Herb and Lois Crisler arrived to shoot pictures for Disney's movie on arctic wildlife, they were new to the Far North. Andy flew them into the Brooks mountains and kept them supplied with food and other necessities. "He flew us to a lonely pass and left us," Lois told me. "We were 150 miles from the nearest humans.

"Next trip, he brought us his personal ivory-trimmed .30-06. Reason? When he left us at the pass, two grizzly bears stood up and swatted at his plane as he took off. It really wasn't his worry if we got into trouble with bears, but he knew we were unarmed. He wanted us to have the protection of a rifle.

"We love Andy," Lois Crisler declared.

As a bush pilot, James L. Anderson was a remarkable, outstanding, giving Alaskan. He worried almost too much about the people he flew into remote places. He regularly flew out of his way to check on lone bush residents, watching for smoke curling from a cabin chimney, fresh tracks in the snow, and other signs of well-being. If all didn't look right he would fret over it until the situation was resolved.

Andy constantly did things for others without asking payment, although serving people in the Koyukuk was his livelihood. I believe that for his personal satisfaction he needed the platform of his wilderness base, from which he could orchestrate the easing of burdens of the mostly Indian inhabitants where there was no Red Cross or Salvation Army.

Crevice Creek Lake is in the southern foothills of the Brooks Range. The white spot in left center is a wall tent used by hunters that I flew to to the lake. JIM REARDEN PHOTO

He was a "do gooder" in the purest sense. Perhaps it was a holdover from his good-deed-daily Boy Scout experiences, or maybe it was backlash from his wartime experiences of dropping bombs on the enemy when his Christian upbringing rebelled at having to kill his fellow man.

Ted Spencer, director of the Alaska Aviation Heritage Museum in Anchorage, believes the term "bush pilot" was first used at Civil Aeronautics Board hearings in 1938 and 1939. Early Alaska pilots called themselves aviators, and some, according to Spencer, despised this new federal designation for their profession. However the appellation "bush pilot" now holds an honored place in our vocabulary and it has spread worldwide. Alaskans generally have great respect and affection for those they call bush pilots.

The daring men who, in the 1920s, flew rickety, short-range airplanes from makeshift runways at Fairbanks, Anchorage, and other northern settlements were, by today's lexicon, bush pilots. By the late 1940s, flying conditions in the Far North had changed. Pilots now had mostly good airstrips, radio communication, top-notch maintenance, and dependable load-hauling aircraft. Here and there in the lonely Arctic a few pilots flew from tiny villages or government-built emergency and refueling airstrips. They landed on gravel bars, unmapped lakes, frozen rivers, snow-covered tundra. They flew miners, trappers, mail, groceries, hardware, sled dogs, sick Indians and Eskimos, occasional hunters—anything and everything.

These pilots were deserving of all the early-day connotations associated with the honorable title of bush pilot.

Andy Anderson was the epitome of such a pilot. He pioneered commercial flying in the Koyukuk country, providing for the first time scheduled air service for a 10,000-square-mile, sparsely settled arctic wilderness.

I discontinued guiding in the early 1960s, and my trips to the Brooks Range diminished. Andy Anderson, after seventeen years of stressful seven-day-a-week bush flying, with an accumulated 32,000 hours in the air, left Alaska to live on his Pennsylvania farm. We seldom communicated.

Then he telephoned. "I'd like you to help me write a book about my flying years," he said.

I paused, about to say no. But hearing Andy's voice brought a flood of memories—wonderful memories—of spectacular mountain flights we had shared, of his many kindnesses, of what he meant to the people of the Koyukuk. I realized Andy's story should be told.

Here it is.

TO ALASKA

In 1946 I was a twenty-four-year-old Navy lieutenant (jg) flying Curtiss dive bombers from the aircraft carrier USS *Princeton.* World War II was over, so I decided to return to civilian life.

During my six years in the Navy, my father had retired from his career as manager of a power plant, and he and my mother bought a 263-acre farm in Pennsylvania. My goal at that time was to become a full-time farmer, so I proposed to pay for repairs to the buildings, fences, and roads at the farm. In return, my parents agreed to turn the farm over to me when they could no longer operate it.

When I arrived home, my father, my older brother Carl, and I decided to start a modern dairy. Dad, in his mid-70s, owned a fine team of horses that he used on the farm. Hoping to modernize, I bought a Ford-Ferguson tractor. My father took one disdainful look at it and bellowed, "Get that damned junk off my property!"

Carl and I tried unsuccessfully to reason with him. We continued to work for a time, but deadlock ensued. The end came when Carl

Two California hunters that I flew into Dall sheep country along the John River

get busy checking out this region of the Brooks Range. JIM REARDEN PHOTO

A Wien DC-3 mainliner sits on the snowy runway at Bettles in the 1950s.
These big planes hauled passengers and freight, serving Bettles three times
a week on their route to Barrow from Fairbanks. ANDY ANDERSON PHOTO

and I found ourselves cultivating—with hoes—a ten-acre corn patch. It was a hot spring day and I was exasperated and sweating as I bent to the endless task, thinking of the marvelous mechanical cultivating equipment parked in our barn. With it we could have finished the job in an hour or so; with hoes we would be at it all day and then some.

I looked at Carl. "Who is crazy here, anyway?"

He shrugged. We loved our father, but his ideas and ours were generations apart. I dropped my hoe, returned to the house, and told Pop, "I've had it. I'm going to go find a job. When you're ready to turn the place over to me, I'll farm it my way. Meanwhile you farm it your way. That way we'll both be happy."

A. C. Barker, owner of Morgantown Flying Service, at Morgantown, West Virginia, hired me as flight instructor and general supervisor. An old-timer in the flying business, Barker owned a J-3 Cub, an Aeronca Champion, a Piper Family Cruiser, an Air Coupe, and a beautifully rebuilt Stinson Junior.

The small plane business was booming, for many recently discharged veterans of World War II were learning to fly on the GI bill. I took exams and flew a check ride with a Civil Aeronautics Administration (CAA) inspector to earn an instructor's license, a cinch after years of intensive Navy flight training. That summer, flying for Barker, I earned three dollars per flight hour while instructing and fifty cents an hour while supervising students' solo flights.

After flying Navy planes with powerful 2,000-horsepower radial engines, the little 65-horsepower engines of the two-passenger Aeronca and other small planes seemed pretty mild. I had to guard against complacency; any airplane can fall out of the sky.

I saw no future with Barker. In August, an ad in a CAA publication caught my eye. "Wanted: Airway controllers for stations in

An eight-story "skyscraper"—the Northward Building—rises behind a log cabin in Fairbanks in the early 1950s. The neighborhood included many old log cabins. JIM REARDEN PHOTO

Alaska. Single men only. Must be able to send and receive Morse code at 40 words a minute."

I was single and knew Morse code. The work was related to flying, the only profession I could claim. It sounded as if it could become a secure job, and Alaska appealed to me. What especially fascinated me was knowing that in the 1920s and 1930s, in crude early planes, pioneering bush pilots blazed trails across this northern wilderness. Alaska, of course, was still a U.S. territory; statehood didn't come until 1959.

I wrote to the CAA. The positive response sent me to Oklahoma, where I spent two months learning CAA radio procedures and becoming a flight controller.

I was ordered to Seattle (by train) and flew to Fairbanks, Alaska, in a Pan American Airways DC-4 Skymaster. This four-engine plane cruised at about 250 miles per hour, and the flight of more than eight hours in that noisy prop plane told me something about the distance between Alaska and the states.

In that year, 1947, the Fairbanks civil airport was Weeks Field, a 3,000-foot-long gravel/dirt runway near the edge of town. It was barely long enough for light bush planes, and not suitable for the Skymaster, which landed at Ladd Field, a nearby Army base. I wasn't dressed for the 30-below November deep freeze I found when I stepped out of the plane, so I didn't stand around.

Downtown Fairbanks was a raw frontier settlement of about 5,500 people. (All of Alaska, one-fifth the size of the continental U.S., then boasted a population of 125,000.) Fairbanks somewhat resembled the towns in old Hollywood horse operas, with their false-front buildings and hitching racks. It was a gold mining and transportation hub near the geographical center of the Territory, 120 miles south of the Arctic Circle. Many streets were unpaved. Log homes were common. The town had two radio stations, two banks, a daily and a weekly newspaper, four hotels, and an abundance of bars. The University of Alaska was three miles away.

Meals and lodging cost half again as much as stateside. Though World War II had ended two years earlier, many Army and Air Force personnel were stationed nearby.

I liked the frontier feeling. I sensed the lighthearted spirit of the friendly, relaxed people. Dress was casual—lots of parkas and furs to cope with the bitter winter cold.

At the CAA office I was directed to a garage where new recruits were bunked. There I found about a dozen men, sprawled on folding cots jammed together so tightly it was necessary to climb into bed from the end. Discouraged, I wondered what I had gotten myself into.

I went sightseeing in Fairbanks, which didn't take long. Snow was piled about a foot deep not only on the ground, but also atop fence posts, power poles, and on trees, giving a Christmassy appearance. The sun swung in a low arc just above the Alaska Range a hundred miles to the south; it didn't produce much heat. Dog teams pulling sleds on the streets only emphasized the distance from my home state of West Virginia.

Smoke spiked straight up from cabin chimneys. Windows of homes and stores were frosted on the inside. Snow crunched with each step. Tires of passing cars made a distinctive squeal on the polished ice. Exotic.

WITHIN A COUPLE OF DAYS I was assigned to Bettles Village, on the east bank of the Koyukuk River 185 miles northwest of Fairbanks and 25 miles within the Arctic Circle. The village was named for Gordon Bettles, who established a trading post there in 1899, and was the farthest north point for the river steamers that traveled the Koyukuk until the 1930s.

The 4,000-foot runway at Bettles was built by Navy Seabees during World War II as an alternate for planes flying between Fairbanks and Barrow, Alaska's most northerly settlement. Wartime traffic included men and equipment for the North Slope of the Brooks Range, where the Navy prospected for oil. There were rumors of large deposits of the black stuff in the region—and of course the huge Prudhoe Bay oil discovery of 1968 confirmed this.

I boarded a CAA Douglas DC-3 loaded with supplies for the hour-and-twenty-minute flight to Bettles Field. As we flew across the lonely land, I peered down at a world different from any I had ever seen. Though it was dark, I was able to catch occasional glimpses of the snowy land. Bettles is surrounded by a vast spruce-birch forest. The great Brooks Range rises gradually to the north. The adjacent winding Koyukuk River and its nearby tributary, the John River, were frozen snow-covered paths oxbowing through the stunted forest.

Bettles Village and the CAA station were five miles south of the runway. The DC-3 landed and taxied to a crawler tractor hitched to a 6-by-20-foot sled. Jim Crouder, a quiet-spoken man, stood beside the idling tractor. Stepping out of the plane was like being immersed in cold water. Air temperature was 40 below.

I was dressed appropriately for a business meeting in Miami. Fortunately, Crouder had an extra parka, which kept me warm from the waist up. Later I wondered if he hadn't said to himself, "Forty below and another damned *cheechako* [newcomer to Alaska]. And dressed in a business suit, would you believe?"

We transferred freight from plane to sled. The DC-3 engines were left running; if stopped they could have cooled too much to restart.

"I'm glad that's done," Crouder exclaimed as he tossed the last box onto the sled. "It's cold enough without having to stand behind those damned turning props."

Jim drove the tractor to Bettles, following the bank of the Koyukuk River. The only way I could keep the lower part of my body from freezing was to walk behind the sled.

By midnight, when the roaring tractor reached the village, I was numb from the waist down. The village included a CAA communications building; a small store in an ancient log building owned by Jim Crouder and his wife, Verree; and two modern houses built by the

CAA with quarters for eight—four unmarried men to a house. Each house had a bathroom, kitchen, living room, and two bedrooms. They were comfortable and unusually modern for that era.

The remainder of the village consisted of a few ancient log cabins inhabited mostly by retired gold prospectors and one Indian family. The only immediate contact with the outside world was through CAA radio. The only transportation link to Fairbanks was provided by the DC-3s of Wien Airlines, which arrived weekly with freight and mail.

The CAA station operated twenty-four hours a day, seven days a week. Each operator worked seven eight-hour shifts. Recreation included snowshoeing, rabbit hunting, card playing, reading, and visiting. With hundreds of miles of deep, snowy forests surrounding the station, snowshoeing was superb. I avoided the high-stakes card games, having lost incessantly in the Navy, but I frequently kibitzed.

Being the newcomer on the block I was a target, the cheechako subjected to Alaskan humor. During a poker game, Frank Theisen, a long-time prospector in his mid-70s, extended an invitation: "Come run my trapline with me tomorrow. It's just a short hike to check my traps and snares."

The snowshoes I borrowed were heavy and awkward as I trudged along behind Frank. The "short hike" along his trapline was five miles—one way. I stuck it out, figuring my relative youth would quickly wear out the old geezer.

Frank broke trail, and at first I rather enjoyed the clean crisp air, the beauty of the wilderness, the blue sky overhead. But after four miles the joy was gone.

"One more snare to check, Andy," Frank encouraged. "After that we'll visit a roadhouse near the last set."

I had visions of hot coffee, the inside warmth of a building, and freedom for a time from heavy snowshoes. We checked the last snare. "Where's the roadhouse?" I asked.

Frank grinned, pointing to a bush with an empty coffee can hanging from it. "Right there, Andy. That's what I call an Alaska roadhouse."

It was my last snowshoe trip with Frank. He was too tough for me, though I had the advantage of being fifty years younger.

KOYUKUK VALLEY RESIDENTS traveled afoot, by dog team, by airplane, and in summer by riverboat. It seemed only natural for me to buy a small plane—a two-place Taylorcraft.

I arranged to stand the midnight CAA shift, which left me sixteen hours each day to sleep, eat, and fly. With long arctic daylight hours between May and September I had plenty of time to fly, and by summer I was spending more time flying than communicating for the CAA.

John Cross, a popular long-time Alaska bush pilot based at Barrow with Wien, inadvertently gave me the break I needed. Flying from Barrow to Fairbanks to have his Piper ski plane serviced, he passed over Bettles, reported his position, and added, "I'm having minor engine problems." He planned to deviate from the regular route to Fairbanks to fly over an area of frozen lakes where, if forced down, he'd have many potential landing sites.

John failed to arrive at Fairbanks. The Bettles CAA station having been his last contact, we played a major role in the search effort. Among the searchers was Fred Goodwin, chief pilot for Wien, who flew a sparkling new ski-equipped Cessna 195. The plane was so new that little was known about its flight characteristics or capabilities. Fred soon located John on the frozen Koyukuk River about twenty miles south of Bettles, where he had landed safely with a dead engine.

Fred eased that beautiful Cessna down in the four- to five-feet-deep, loose, dry, powdery snow near John's crippled Piper. The aluminum plane plowed deeper and deeper. Snow flew high, almost concealing the submerging craft from John's view. When the Cessna stopped, only its top half was visible. About then Fred realized his 2,030-pound airplane was a bit heavy for ski work in such conditions.

Now two pilots and planes were marooned on the Koyukuk River.

Next morning I talked with Fred by radio, promising after I got off watch to help snowshoe a runway for the Cessna. With extra snowshoes and a snow shovel I flew to where the two partially buried planes lay. Circling, I saw the two pilots hadn't accomplished much in the way of preparing a runway.

I touched down gingerly, keeping my airspeed high while dragging the skis on the snow's surface. I fed in more throttle and lifted off to circle again, then touched down for a full stop in the tracks I had left. My 950-pound T-Craft poised itself nicely atop the snow.

Neither pilot had eaten for some time. I flew back to Bettles for food. As I took off, they were starting to pack a snow runway for the Cessna.

At Bettles I scrounged for food and noticed that Johnny Musser, my roommate, had baked an apple pie. It was just the thing for a couple of hungry pilots.

"Can I take half of this pie?" I asked.

"Sure," he said.

As I tried to cut the pie in two, my knife encountered hard objects. Johnny had used dried apples in the pie.

"Johnny, did you cook the apples before putting them in the pie?" I asked.

"No. Are you supposed to?"

I returned to the stranded pilots (sans pie) with enough food to satisfy. All day the three of us compacted the snow into a runway by snowshoeing repeatedly in front of the Cessna. Just before dark Fred attempted a takeoff. He was still struggling to get into the air at the end of our painfully packed strip when his plane again plunged into deep, untouched snow. For moments that aluminum beauty was completely hidden in whiteness. When the snow settled, John and I looked at each other, shook our heads, both rolling our eyes skyward.

"Damn it," he said. "We're right back where we started."

We called it a day. I flew John and Fred to Bettles in the T-Craft. They both ate like starved wolves and enjoyed a good night's sleep.

The following morning, while flying Fred back to his buried plane, he asked if I'd be interested in working for Wien. "Sure," I said, "provided that Wien establishes Bettles as a distribution point for freight, mail, and passengers, and I can remain here."

Federal regulations prohibited commercial flying by CAA employees. "Conflict of interest," they called it, but I paid scant attention.

After I had been flying from Bettles for several months, including a number of flights for which I received payment, the CAA regional administrator warned me, "Andy, you can work for us and you can fly for yourself. But you cannot work for us and fly commercially."

Fred Goodwin's offer and the edict from the CAA came at about the same time. It wasn't a hard choice to make. I could earn more in a day of flying than I made in a month with the CAA. I came to an agreement with Wien and without regret resigned my CAA position June 1, 1948.

That summer I flew passengers from village to village along the 554-mile-long Koyukuk River, Alaska's third-longest. Emergency trips to hospitals at Tanana and Fairbanks became almost routine. Flying my T-Craft on floats, I took many fishing parties from Bettles

Field into the myriad surrounding lakes and streams. The Brooks Range region was then virtually unfished. The area's clear lakes and streams abounded in grayling, lake trout, and Dolly Varden; schools of arm-long pike in this sportsman's paradise had never seen a lure.

That fall I flew hunters into lakes and streams in the foothills and peaks of the Brooks Range. They found an abundance of moose, caribou, white Dall sheep, and both black and grizzly bears.

It was the start of my seventeen-year career flying from Bettles Field as an arctic bush pilot.

BATTLESHIP SAILOR

I grew up in the small coal mining community of Montcoal, West Virginia. My father, Fleming "Fred" Anderson, ran the plant that produced electricity for the mine and town. The community included a church, school through twelfth grade, a YMCA, health clinic, a bathhouse for the miners, and a general store.

It was a splendid town for a growing boy, with endless job opportunities. I cut wood, cleaned yards, mowed grass, washed cars, pruned bushes, painted fences, and had a daily paper route. When in high school I was janitor at the local store, and later clerked there.

I once lived for six months with my grandmother Anderson in her four-room log cabin on a 120-acre farm near Alderson, West Virginia. There I developed a yearning to live the life of a farmer. The Great Depression was in full swing and jobs were scarce. Although barely out of short pants, I sensed that life was uncertain and decided that if I owned a farm I could always raise enough food to support me and mine. From that time having a farm became my lifelong goal.

I was a radioman, second class, aboard the battleship USS North Carolina *when this photo was taken in 1944.*

In Montcoal I became a Boy Scout and obsessively earned merit badges and within a few years became an Eagle Scout.

I greatly admired my high school biology teacher and decided to excel in her class. At the expense of other subjects I concentrated on biology. Despite this effort, my grade at the end of a test period was only C.

I wanted to get my teacher's attention, so I dropped the course, expecting she would talk me into coming back. I needed biology to graduate and couldn't imagine she would allow me to quit. But she did, and the rest of my class graduated from high school without me.

I was too ashamed to attend the exercises. When I returned home late, my mother, Mary Ann, had already gone to bed. I went to her room to bid her good night and received a lesson of a lifetime.

"There's a gift on the mantle for you," she said.

It was a beautiful wristwatch. I thanked her for it, full of shame at letting her down.

"Turn it over and read the inscription," she said.

On the back of the watch I read, "For the effort that failed."

I have never forgotten that message. Or my shame. I vowed then that somehow I would make it up to her.

After I left high school my father hired me to work with him in the power plant. That summer of 1941 my odd jobs and saving ways paid off when I bought my first new car, a $995 four-door Chevrolet Caprice.

As a boy I was nuts over airplanes. On the rare occasion when a plane flew over Montcoal, we kids cried "Airplane, airplane" as we pointed up. Occasionally a barnstormer dropped leaflets announcing where he would offer rides from a farmer's field, for there were no airports.

My only plane ride came at the Lewisburg County Fair when I

rode for about fifteen minutes in a stagger-wing cabin model with a radial engine. The window seat I was strapped into didn't offer much of a view.

JAPAN'S SNEAK ATTACK ON PEARL HARBOR on December 7, 1941, changed forever the lives of millions, including mine. Five days later my brother Carl and I enlisted in the Navy.

At boot camp at Norfolk, Virginia, a tubercular spot was detected on one of Carl's lungs. He spent several years recovering, and he eventually became a federal bank examiner. He died in 1997—regretting to the last that he was unable to serve his country in war.

My boyhood dreams of becoming a pilot had been with little hope, for money was short. Once in the Navy, I determined to become a naval aviator. However, I didn't have the required high school diploma.

Boot camp ended abruptly when four hundred of us were sent by train to Miami and then bused to the Naval Air Station at Key West, where we boarded two destroyers.

"No lights, no talking," was the order. We were told we would soon board another ship at sea. Hours later whaleboats were lowered to relay us to a huge ship looming out of the darkness. I climbed a rope ladder hanging from the deck with my eighty-pound seabag over my shoulder.

Only when I stood on the steel deck of that behemoth did I learn I was a crew member of the USS *North Carolina,* our newest battleship, commissioned in 1941. The 729-foot-long ship carried nine 16-inch guns, each capable of tossing a 2,000-pound shell twenty miles or more, plus many smaller guns.

I requested assignment to the aviation division. "Sorry," the personnel officer said. "That division is full. What did you do in civilian life?"

"Water tender and fireman at a coal mine power plant."

"Fine," he said. "You're now in the engineering division of the *North Carolina.*"

My new life included frequent engineering training classes. Off duty I associated with men in the aircraft division. I learned that most of these men had started as a radioman/gunner and that Morse code was a requirement.

I successfully applied for transfer to the communications division, where I quickly learned radio procedures and Morse code. I was promoted from seaman to radioman third class, always watching for an opening in the aviation division.

Our ship sailed through the Panama Canal and went directly to the South Pacific war zone. It was early August 1942, and the Navy had built enough strength to challenge the Japanese at Guadalcanal. The camouflage-painted *North Carolina's* main duty was to protect aircraft carriers from air attack, for we had enough firepower to plaster the skies.

Our first battle lasted several dismal days. Our support aircraft were shot out of the sky by the numbers. On August 8, in the Battle of Savo Island, one of the Navy's worst ever, Japanese cruisers sank our cruisers *Quincy, Vincennes,* and *Astoria* and the Australian cruiser *Canberra.*

On September 15, the *North Carolina* was patrolling off Guadalcanal, accompanying the carriers *Wasp, Hornet, Saratoga,* and *Enterprise,* as well as heavy cruisers and destroyers. I had worked the 4 A.M. to 8 A.M. shift in the radio shack. Sound asleep in my bunk, I was awakened by a tremendous blast. The ship rolled violently, tossing me to the deck. The men in my compartment were shocked and frightened. Lights remained on, but the compartment filled with smoke.

"We've gotta get out of here," one man shouted.

A barefoot sailor in skivvies, carrying his pants, headed toward the hatch. Too late. All hatches had already been closed and sealed by other sailors following damage-control procedures. The sealed hatch blocked me from reaching my battle station at an antiaircraft gun.

The ship had not received a fatal blow, and hatches were soon reopened. A Japanese submarine, the I-19, had fired a spread of six Long Lance torpedoes at the carrier *Wasp.* Three slammed into the *Wasp,* sinking the ship. The other three streamed seven miles farther to the task force centered around the carrier *Hornet.*

One of these torpedoes struck the *North Carolina* near the forward turret, tearing an 18-by-32-foot hole twenty feet below waterline and killing three men. Another slammed into the port bow of the destroyer *O'Brien,* causing damage from which it later sank. The attack was probably a record no submarine has ever equaled: three warships, two them nearly seven miles from the target, struck by five of six torpedoes in a single spread.

WE SAILED TO PEARL HARBOR for repairs. At this time I was finally admitted to the aviation division.

The *North Carolina* carried two Vought Kingfisher OS2U airplanes. On the ship, each plane was mounted on a sled, which in turn was mounted on a sixty-foot-long catapult. The planes were catapulted into the air.

Ruggedly built and powered by a 450-horsepower Pratt & Whitney radial engine, this observation plane carried a pilot and a crew member, a 100-pound bomb under each wing, and five hundred rounds of .30-caliber ammunition. Top cruise speed was 140 miles an hour.

I have a vivid memory of the first time I was catapulted in a Kingfisher. The canopy was open, and the pilot and I wore life jackets in the very likely possibility the Kingfisher splashed into the sea.

The pilot set flaps, centered the control stick, and anchored his right elbow in the pit of his stomach so the stick wouldn't move at blastoff. He shoved the throttle wide open, locking his hand against it to keep it from slipping in the violent pushoff. The plane vibrated as 450 horses plunged.

I sat uneasily in the rear, yet eager, tensely braced. The roar of the engine was almost overwhelming. My pilot nodded. Black powder in a piston at the base of the sled was fired. With a loud *boom,* smoke blossomed as the Kingfisher accelerated to 75 miles an hour within sixty feet.

My cheeks, belly, and legs were momentarily jerked violently back as the plane leaped. Heavily loaded, we sank below deck level before the prop and wings caught enough air to climb.

Once aloft I turned around and rode backward, manning the twin swiveling .30-caliber machine guns, scanning the sky for enemy planes. I also worked the radio and signal lamp.

Kingfishers were the ship's eyes. The *North Carolina* could be twenty miles from land targets or enemy ships while it hurled its 16-inch shells. We reported where they landed so gunners could correct their aim.

We searched for lurking submarines. The bombs we carried were for use against such raiders. Radio silence was kept on most flights; I communicated with the ship with a powerful signal light.

To land the Kingfisher, the speeding *North Carolina* made a

270-degree turn to head directly into the wind. The ship's speed combined with the turn created a slick where the plane, equipped with a float, landed. The plane then taxied rapidly, sometimes at 25 miles an hour or more, moving onto a rope sled dragged from the ship's stern. A hook on the bottom of the plane's float caught the sled when the plane slowed.

I crawled out on the tippy wing to capture a hook lowered from the ship by a crane and inserted it into a slot; the crane then hoisted the plane aboard.

Through all this action a dark cloud hovered over my goal to be appointed to Navy flight school: the required high school diploma.

My mother told my former high school principal of my dilemma. He solved my problem by giving my mother a diploma bearing my name, commenting, "I think by now James has earned this."

While Pearl Harbor repairs were being made on the *North Carolina*, I took the physical exam for flight school. "Sorry," the doctor told me, "you're underweight."

I stuffed myself with bananas and everything else I could find to eat. I then returned to the doctor. "Sorry, Anderson," he said. "We want honest weight, not weight from eating bananas for a week."

Few members of the *North Carolina* crew were plump. Shipboard food was OK, but waiting for a torpedo, a bomb, or enemy shellfire was a constant mental strain; tropical heat and humidity didn't help appetites.

The ship returned to the war zone. While we were anchored at New Caledonia I learned a flight surgeon was aboard a nearby hospital ship. I reported to him and requested a physical exam—and passed.

Weeks went by and nothing seemed to be happening with my application. On one of the ship's visits to Pearl Harbor, I thought of

Commander Hill, our one-time gunnery officer, now on the staff of Admiral Nimitz.

Deciding to go over the heads of my shipboard officers, I marched into Admiral Nimitz's domain and asked to see Commander Hill.

"And what do you want to see him about?" his yeoman asked.

"It's personal, and important. I once served under him," I said, stretching the truth.

My legs felt like rubber when I walked into Commander Hill's office. Fortunately he remembered me and allowed me to explain my problem. "I knew if anyone could help, it would be you," I stammered.

He told me to check later with his yeoman.

Two days later his yeoman asked for my name and service number. In a file he found my card, and with a sly wink he moved it from rear to front.

Two weeks later I received orders to return to the states to enter the University of Texas as a Naval Air Cadet.

Today the USS *North Carolina* is on permanent exhibit at Wilmington, North Carolina. I visited her a few years ago and found my old bunk and other familiar compartments. As memories of one-time shipmates and wartime events returned, I had to flee ashore before tears began to flow.

CHAPTER THREE

NAVY PILOT

I had to complete class work at the University of Texas at Austin before learning to fly. Classes lasted from 6 A.M. to 10 P.M. I had to memorize much of the curriculum, finally realizing how I had foolishly squandered precious high school years. "Drop your pencil and pick it up and you're behind two weeks," one student ruefully commented.

I struggled in navigation class. The turning point came when we were given a problem requiring students to make a change of course midway on a theoretical flight.

The instructor came into class next day with a frown as he dropped the corrected exam papers on his desk. "Only one student understood this problem. The rest of you ran out of gas and ditched in midocean."

To my surprise he called, "Anderson, stand up. You were the only one to solve the problem."

That gave me confidence, and I sailed through the rest of the course without difficulty.

Ten out of the twenty of us who started the course passed and were transferred to Baylor University at Waco, Texas, where we

Members of Flight No. 347, Naval Air Station, Jacksonville, Florida, pose with a Navy SB2C Curtiss Helldiver. I flew these dive bombers in combat in the Pacific during World War II. In this photo, I'm on the far right in the rear row.

U.S. NAVY PHOTO

learned to fly the three-seat Piper J-5B Cruiser, a fabric-covered plane powered by a 75-horsepower Continental engine. The student sat in the front while the instructor occupied the wide two-person seat in the rear.

My instructor taught me how to prepare for a flight by making sure the fuel tanks were full, that there was sufficient oil, and that all bolts and other fittings were tight and locked. He flew us to a practice area and had me take over. My first exercise was to learn how to coordinate turns.

Under my control the plane skidded or lost and gained altitude. I overcompensated and threw us all over the sky. The instructor was incredibly patient. He spoke calmly in my right ear as I tried to follow directions.

After an hour he landed the plane and gave me advice that helped throughout all the instruction to come. "Sit in a quiet place and think your way through what we have just done. In your imagination refly the airplane, moving the controls in your mind. Do it repeatedly."

Next flight went better. Like most students, however, I tried to force the plane to do what I wanted it to do. When my instructor judged I had struggled long enough, he spoke the magic words: "Anderson, let the airplane fly itself. All you do is point it in the direction you want it to go."

He trimmed the plane so it flew hands off. "Take your hands and feet off the controls. Don't touch a thing," he ordered.

I complied reluctantly. The pilotless plane flew a series of up and down movements and gradually stabilized itself into straight and level, effortless flight.

Flying suddenly became easier. Next he taught me how to land. I bounced a few landings, and it took a few hours to learn how to keep the plane headed properly with the rudder, which is operated by foot pedals.

I soloed after eight and a half hours of instruction. I had to take off, climb to 600 feet, fly the landing pattern, and land—three times. My instructor stood on the ground watching, as flight instructors have traditionally done since the first airplanes. No doubt he also followed tradition by gritting his teeth and praying I had learned enough to return to earth gently.

My first two takeoffs and landings went well, leaving me grinning. I was now a pilot! I could fly! It was a marvelous feeling no pilot ever forgets.

On my third attempt, on the downwind leg I pulled on the throttle when the plane was opposite the point where I wanted to touch down. The throttle refused to move. I pulled harder. It still refused to move. The engine continued to drone at 2,200 rpm instead of the 1,500 prescribed for a landing approach. The throttle remained stuck no matter how hard I yanked.

I decided to stop the engine and land dead stick—with the propeller stopped. On final approach when I judged the plane would glide safely for the remaining distance, I pulled the mixture control, stopping the flow of fuel to the engine. The engine stopped, and I dropped the nose and glided to a gentle landing.

I was elated. I had coped with an in-flight emergency. And I had successfully taken off and landed three times. But no sooner had the plane coasted to a stop when my instructor shoved his head through the door, "What in hell do you think you're doing, Anderson? You don't grandstand on your first solo flight."

"The throttle stuck!" I exclaimed

"Bulloney!"

"Try it yourself," I said.

During a rest period on Saipan Island between battles, I posed for this photo with part of my crew. I was squadron maintenance officer aboard the aircraft carrier USS Princeton. *I'm third from left in the front row of this group next to a Curtiss Helldiver.* U.S. NAVY PHOTO

51

He climbed in and started the engine. I almost laughed when the engine roared at full throttle and he had to jam on the brakes to keep from taxiing wildly across the pasture. He tugged at the throttle, but it remained stuck.

Sheepishly he endorsed my three-landing solo flight. I now had training roughly equivalent to that of a low-time private pilot.

IN TIME I ARRIVED at the Memphis Naval Air Training Center, Tennessee, one of the largest wartime bases in the U.S. If a student pilot's skills didn't meet standards, he was out, no exceptions. Washouts were common.

We flew the Stearman biplane. We called it "The Yellow Peril" for its color—and perhaps because those who flew it were novice pilots. Powered by a 220-horsepower Continental radial engine, this rugged 2,700-pound plane could take about any kind of aerial punishment.

The two-cockpit Stearman is easy to fly. The student rode the front cockpit, the instructor the rear. Communication was via a gosport tube, a one-way instructor-to-student link. The student wore earphones, the instructor had a mouthpiece. There was no student back-talk.

Because it is a biplane with an abundance of wires and struts and a big blunt engine, when power is reduced it drops like the proverbial brick. The greatest weakness from the student pilot's standpoint was the narrow landing gear; crosswind landings held any pilot's atten-tion. There were no flaps to slow landings.

My naval officer instructor spent hours explaining how and why an airplane flew. With his skilled teaching both aloft and on the ground, I advanced steadily. I had to master ordinary maneuvers, precision aerobatics, formation flying, emergency landings, landings to a circle, night flying, instrument flying, and cross-country flight. If a pilot dinged an airplane during any of these exercises he was automatically washed out.

Landing to a circle almost did me in. I was required to fly a downwind leg at exactly 600 feet, and when opposite a fifty-foot circle on the landing field, I had to reduce power, turn, and in three out of five attempts land the plane within the circle.

Two landings went nicely. I was short on the third and stalled the plane a few feet from the ground. It dropped, landed on the left wheel, and started to turn sharply left with the left wingtip dangerously close to the ground. Somehow I couldn't get the right wheel down. Frantically I shoved in full throttle and the plane left the ground while still running on the left wheel. Fortunately the wingtip didn't touch.

"If you had scraped that wing it would have been back to the fleet for you," my instructor warned.

Flight instruction was always interesting. Once I was making night landings with about ten other Stearmans in the pattern. In turn we climbed to 300 feet, turned left while still climbing, and at 600 feet set up for another landing. All went well until one of the planes started making right turns. After taking off, that errant plane returned to land after four right turns.

Instructors, on the ground, were perplexed. One plane was flying a right pattern; nine were flying the prescribed left. The Stearmans were not radio-equipped. Traffic was controlled with lights; green signaled a pilot to land, red was a wave-off.

The instructors recalled all planes. The guy making right turns was a French student with a minimal command of English; many trainees from allied countries were sent to the U.S. to learn to fly.

"Why were you turning right when the pattern was left?" he was asked.

"This airplane won't turn left," he responded.

At this stunning announcement the instructors inspected his plane with flashlights. Sure enough, his plane would not turn left. They quickly discovered why.

On the ground, it was the practice of line crews to attach a 2-by-4 board on edge on top of the bottom wings of each plane—spoiling airflow and keeping the planes in place when winds picked up. In preparing for his flight, the French student had failed to remove the 2-by-4 from the right bottom wing.

At last I had only to complete a cross-country flight to earn my gold wings. A friend who had failed the program received permission from the base commander to accompany me.

My friend rode the front cockpit and was eager to fly. I gave him the compass course to follow to an Arkansas Navy base and relaxed, paying no attention to our progress. When I thought we should have been near our destination, none of the landmarks corresponded with my map.

I casually landed in a field and asked a farmer where we were. He laughed and showed me on the map that we were miles off course. I then understood why my friend had flunked as a cadet.

Preparing to take off, I suddenly realized that ponds blocked each end of the field. Beyond the pond in the direction of takeoff was a fence. Above the fence were high-tension wires. The field was sodden from heavy rain, and the plane's wheels had sunk deeply.

I was in trouble. My Navy career hung in the balance.

I put the tail wheel in the far pond, held the brakes, and shoved the throttle until the engine was wide open. I released the brakes and we slowly gained speed. I jerked us into the air. We were going to make it over the fence, but I was still trying to climb above those deadly high-tension wires. We weren't going to make it. I pushed the stick forward and flew under the wires and over the fence.

I was awarded the much-sought-after Navy wings of gold, which I wore with great pride. I was promoted to aviation pilot first class, an enlisted rating.

The names of about five thousand enlisted naval pilots of World War II are on record at the Naval Air Station museum at Pensacola, Florida. My name is included, a distinction I'm very proud of.

I WAS ASSIGNED to the Naval Air Station at Corpus Christi, Texas, where I flew Navy mail and personnel throughout the U.S., mostly in a twin Beechcraft D-18. This retractable-gear transport, with two 330-horsepower radial engines, cruised at more than 200 miles per hour.

Occasionally I test-flew planes that had been worked on before they were returned to flight status. One morning the maintenance officer asked me to test-fly an SNJ, a low-wing, 550-horsepower, two-seat basic trainer.

The plane had been overhauled and control cables replaced. I preflighted it, climbed in, started the engine, and called for taxi instructions. At the end of the runway I checked magnetos, propeller, and control movement. All appeared normal.

The tower cleared me and I advanced throttle and the plane rolled. As it gained speed I pushed forward on the controls to raise the tail. To my surprise the airplane leaped into the air and I was suddenly flying. I knew instantly something was wrong with the controls.

I was about thirty feet in the air and didn't dare do anything drastic. I carefully experimented. Normally when the wheel or stick of a plane is pushed forward the nose goes down. On this airplane, when I pushed the stick forward the nose went up. I almost couldn't believe it. To myself I said, "Let's try this again, cautiously."

It was the only time I ever climbed an airplane by pushing forward on controls. Fortunately the aileron control cables were OK and the plane banked left when I moved controls left. I radioed for clearance and started to school myself on moves I needed for landing.

It went against everything I had learned to work that reversed elevator control. I held the plane off for some distance and allowed it to settle gently onto the runway by holding forward pressure (normal is to hold back pressure). It was one of my better landings.

At the hangar, the maintenance officer looked at me questioningly and I grinned, partly in relief at having landed safely and partly because I had something on him.

"The elevator controls are backward," I said.

He rushed to the plane and worked the controls. Like me, he was disbelieving. He turned to me in great distress. "You could have been killed."

"Yeah, but I wasn't," I said.

And you could have been court-martialed for this, I wanted to say, but didn't.

"I won't say anything about this if you don't," I said.

Part of the fault was mine; I should have caught the improper elevator movement when I checked controls before takeoff.

Later I test-flew the plane again—after very carefully checking the controls. It flew beautifully.

My duty at Corpus Christi ended abruptly because the Japanese were shooting down many of our planes and pilots, and more combat pilots were needed. Many enlisted pilots were hurriedly commissioned as ensigns and reassigned to the fleet or to further training.

I became an ensign and was sent to Cecil Field at Jacksonville, Florida, to learn to fly the SB2C Curtiss Helldiver—a dive bomber.

THE HELLDIVER WAS BIG, bulky, and weighed a whopping 10,114 pounds. It wasn't an airplane to caress; a lot of rudder was required to get a response, and the rudder was oversize.

We called it The Big-Tailed Beast, or simply The Beast. We twisted the official designation of SB2C into Son-of-a-Bitch, Second Class.

To overcome the plane's tendency to recover too abruptly from a dive, Curtiss built a fifteen-pound counterweight on the bottom of the control stick. To give it credit, the plane was rugged and could withstand a lot of Gs in recovering from a dive.

Its 2,000-horsepower Pratt & Whitney radial engine swung a thirteen-foot, three-blade propeller that yanked the plane into the air with a torpedo hung in the bomb bay and a 500-pound bomb under each wing. We often had four rockets slung under each wing. Cruise speed fully loaded was about 160 miles per hour.

My Cecil Field instructor was never satisfied until I could place a bomb within a fifteen-foot circle after starting a dive at 10,000 feet or higher. A pilot was expected to go as low as necessary to score a hit.

I was assigned to the new Essex-class aircraft carrier USS *Princeton*, namesake of the original *Princeton* sunk by a Japanese dive-bomber during the Battle for Leyte Gulf. I joined the ship in March, 1945, at Guam when preparations were being made for the invasion of Okinawa. The Pacific war was winding down as we built our military strength for the invasion of Japan.

I was assigned to Squadron VB 81, which had about one hundred pilots flying Corsair F4U fighters, Corsair F4U fighter bombers,

Curtiss Helldivers, and torpedo bombers. We could usually put together a strike force of about eighty planes.

My squadron commander expected his pilots to follow his aggressive example. Sometimes when I was fortunate enough to return from a mission, his exhortations of just how damn stupid I was made me almost wish I had been shot down.

On missions we usually left the *Princeton* about 4 A.M. and climbed over our target, usually about two hundred miles from the ship. Fighter Corsairs went in first to draw small-arms fire. Fighter bombers followed to create confusion and soften the target. Dive bombers started their plunge from 10,000 feet or higher. We peeled off from formation every fifteen seconds and it wasn't unusual for me to see six or more planes diving ahead of me. Dive flaps kept our speed at about 275 miles an hour.

After dropping bombs we flew low and ran like hell. Smart Japanese gunners shot into the water ahead of our planes, trying to produce geysers to knock us down.

Taking off and landing on the carrier was routine, although some spectacular crashes occurred. The secret to successfully landing on a carrier lies with the Landing Signal Officer (LSO) who directs pilots with signal paddles that he holds in his hands. His signals are mandatory; the pilot, in puppet fashion, merely moves controls at his order.

The squadron once returned from a mission to find a tropical storm pounding the ship. The *Princeton* was almost standing on end as it climbed gigantic waves, then plunged into troughs. I flew into the touchdown area where the LSO took over. It was his responsibility to maneuver my plane so at the moment the carrier deck rose to the top of a wave and started to descend, he could signal for me to cut the engine. The plane was supposed to hit as the deck fell in its roller-coaster plunge.

I arrived when the ship was at the bottom of a wave and had started up. The LSO signaled me to cut engine. My plane fell uncontrollably and violently met the swiftly rising deck. I could do nothing to cushion the impact.

That big Curtiss hit so hard the engine mount buckled and both wings drooped. My head was forced between my legs. The flight surgeon leaped on a wing, grabbed me by the hair, and jerked me upright.

"Are you OK?" he yelled.

"Yeah," I said, dazed.

Sure I was OK, except for my ego.

The deck crew yanked me and my radioman/gunner out of the plane and pushed the ruined Helldiver overboard.

I was still aboard the *Princeton* when the war ended, and like every military man in the Pacific, gave thanks for the atomic bombs that provided that ending. I decided on a trial year in the peacetime Navy.

During that year the ship was in China, where antagonism between Russia and China bordered on a potential third world war. Occasional minor military clashes flared. I thought our orders to not shoot back if shot at were ridiculous, and was always uneasy on patrol flights, feeling I was a sitting duck.

Aircraft maintenance slumped because many mechanics had left. Peacetime promotions came slowly. Training and flight times were cut, though pilots were expected to maintain wartime proficiency. It was a discouraging outlook.

In May 1947, I returned to my parents' farm in Pennsylvania. I had served in the Navy for six years, starting as an apprentice seaman and leaving as a lieutenant (jg). I participated in four major battles, but my memory of those battles is hazy. It's probably just as well. War is hideous.

THE T-CRAFT AND THE CHEECHAKO

I was nicely settled in at Bettles by early 1948 when I bought a year-old Taylorcraft airplane. I got it for $3250 from Jim Magoffin, owner and operator of Interior Airways at Fairbanks. The T-Craft, built primarily as a trainer, was designed to handle skis, wheels, and floats, but it didn't have a starter, generator, or battery. Omission of this equipment gained about fifty pounds of payload, but the prop had to be pulled through to start the engine.

The 65-horsepower engine had been modified to produce 75 horsepower. A canvas sling had been installed behind the seats—I called it my "Alaska baggage compartment." I could safely fly three hundred pounds in addition to myself, either as passengers, freight, or in combination. It had space for the pilot and one passenger in side-by-side seating. I have fond memories of that light-blue airplane. It was forgiving and easy to fly. With it I learned much about flying the arctic Alaska skies.

With August snows on the Brooks Range, about 1960, I taxied this Cessna 180 on Agiak Lake. I was serving a hunting party I had dropped off there. JIM REARDEN PHOTO

A dead calm settles on Crevice Creek Lake, on the south slope of the Brooks Range. While they don't have the towering heights of the Alaska Range peaks, the mountains of the arctic Brooks Range are among the most scenic in the world. JIM REARDEN PHOTO

When equipped with floats, the plane could be landed on a small circular lake too small to take off straight across. In taking off I would make a steep left turn, lift the right float out of the water, and continue to circle the lake in that attitude until I gained enough speed to lift off.

When Jim Magoffin delivered the T-Craft to me at Bettles Village, I parked it on the frozen Koyukuk River. To heat it, I ran an extension cord from the CAA generator and made shade bonnets for 75-watt light bulbs from two five-pound coffee cans. The cans kept the lights away from vulnerable engine wiring and other heat-sensitive parts. With a bulb on each side of the engine, and the cowling wrapped with a canvas engine cover, the engine was always warm. On mild days of 15 to 20 below zero. I had to remove one of the lights; otherwise the engine became too warm.

At first I didn't fully realize what that little T-Craft was going to mean to me; I thought it was simply a way for me to get around on my own. I could fly to Fairbanks or to villages in the Koyukuk region, where I started to become acquainted with the mostly Athapaskan Indian residents, Eskimos, and the white traders, teachers, and missionaries. I started to learn the geography of this great wilderness region and after a time I felt comfortable flying over it. Owning that little airplane gave me an extraordinary feeling of freedom. Eventually, owning the T-Craft led to what turned out to be the most important work of my life.

Among the wonderful people I became acquainted with when I arrived at Bettles Village were Warren C. "Canuck" Killen, who was from Canada, and his partner Jack Irwin. They worked together on construction projects at Fairbanks and had decided to branch out on their own. In the fall of 1947 they loaded an outboard-powered riverboat with their belongings and headed down the Tanana River to the Yukon, then down the Yukon to where it is joined by the Koyukuk.

They then motored upstream more than four hundred miles on the Koyukuk to Bettles Village.

Shortly after arriving at Bettles, Canuck started courting Nellie Withrow, a local girl. By the time I acquired my T-Craft, they were serious about each other and soon decided to get married. They wanted U.S. Commissioner Charlie Irish at the early gold rush settlement of Wiseman to tie the knot.

They offered to pay me to fly them the sixty miles to Wiseman. I was agreeable, but there was a problem: My T-Craft seated only two. But after I added the combined weights of Canuck, Nellie, and me, I decided the plane could handle us. I proposed that Nellie sit on Canuck's lap for the forty-five-minute flight. She was willing, so away we went. Charlie tied the knot, and the three of us returned to Bettles, with Nellie still on Canuck's lap. It was a cozy wedding trip—and my first commercial charter in Alaska.

In 1948 I was still a cheechako with a great deal to learn about the Arctic. From the beginning I tried to expand my operational range, but the aviation maps of the time were sketchy and I couldn't trust them. I was cautious about flying into unknown country, for likely there would be no other person within hundreds of miles if I was forced down.

My winter flying was on skis, enabling me to reach an almost unlimited number of snow-covered lakes, rivers, and flats. My summer flying was with the plane equipped with floats or pontoons. For these, there were endless miles of river and hundreds of lakes.

Local rivers were my landmarks. In time I learned to memorize landmarks for all four seasons, for seasonal changes can drastically alter the appearance of objects on the ground. I eventually became familiar with landmarks in the major river valleys—the Tanana, Yukon, Koyukuk, Kobuk.

I was surprised to find that people who had lived in the region for years commonly didn't recognize major landmarks from the air. It took me a couple of years before I could spot landmarks wherever I flew within a radius of more than five hundred miles.

As I started to fly commercially—to haul paying passengers and light freight—I purchased aviation gasoline in Fairbanks in fifty-gallon drums which were airfreighted to Bettles Field thirty-five drums at a time. By the time I burned it in my little T-Craft, that gasoline cost about eighty cents a gallon. The plane burned about four gallons an hour as it cruised at 90 miles an hour. I set twenty-five dollars an hour as my charter rate.

The metal skis that came with the plane commonly froze down when the plane was parked, and it was necessary to rock the wings or to have someone push the plane to break it loose. That winter I called my operation "Jim Anderson's You Push 'Til We Get Started Airline."

Shirley English, wife of Wien Airlines pilot Bill English, asked me to fly her and their little baby from Bettles to Wiseman, where she was to visit Bill's mother. It was very cold as Shirley sat with the baby in her arms while I revved the engine, trying to move the plane. The skis were stuck fast. I idled the engine, climbed out, and rocked the wings. Those damned skis remained locked in place. I then pushed and pulled on the tail, without success.

Finally I said, "Shirley, you're going to have to help me. I'll hold the baby and give the plane full power while you get out and push on the tail to break us loose."

She did, I did, and the plane broke loose. And though half a century has passed, Bill English still likes to remind me of that embarrassing incident.

Noel Wien, one of the airline's founders, once gave me a tip on flying with metal skis. "Spread a little kerosene on the snow in front of metal skis," he said. "When they break loose, they'll move easier." It works, but first you have to break the darned things loose.

In the early years of winter flying, planes used wooden skis. Then came the pesky metal ones. Next came ski/wheels, in which the wheels could be hydraulically raised or lowered to protrude slightly below an opening in the skis. This largely solved the problem of freezing down.

DURING THE FIRST WINTER, I spent much time with Canuck and Jack Irwin, frequently accompanying them when they cut firewood and hauled it to their cabin with a dog team. This was always an adventure. I love dogs and it was fun to see how much the dogs enjoyed the trail.

On one wood-cutting expedition Jack cinched the bindings on one of his snowshoes so tightly it restricted circulation and he froze his foot. In a few days gangrene set in and he urgently needed medical care. He asked me to fly him to Fairbanks where he could go to St. Joseph's Hospital.

My daily watch hours with the CAA were from midnight until 8 A.M. On the day I flew Jack to the hospital, I lifted the T-Craft off the frozen Koyukuk River half an hour after I got off duty. I expected the flight to take two to two and a half hours each way. It was late winter, with less than six hours of daylight.

The flight was beyond the area I had previously flown, and I was unfamiliar with the route. Jack assured me that once we reached the halfway point at the Yukon River, he would recognize the east-west Livengood Highway, which we could follow into Fairbanks.

When the sun rose and started to follow its low arc above the horizon, it proved to be a beautiful, clear day. However it was cold;

nighttime Yukon Valley temperatures had plunged to 60 and 70 below zero.

As we flew along at 5,000 feet I remembered the owner of a T-Craft, a twin to mine, who had recently made a flight from the high-country Brooks Range, bound for Fairbanks. His route should have taken him over the village of Fort Yukon on the Yukon River. He must have missed Fort Yukon and then flown up and down the Yukon River trying to determine his location. With his fuel low, he apparently spotted a trapper's cabin, circled, and found a place to land some distance away.

In extreme cold it is life-threatening to perspire heavily. Damp clothing allows cold to penetrate, hypothermia can set in, and death follows. That may have been what happened to this pilot. He was found dead, frozen solid, lying along the trail between his plane and the cabin he was trying to reach.

I was thinking of this as Jack and I enjoyed the spectacular view of the magnificent land below. As planned, we hit the Yukon River directly over Stevens Village. Soon after, I spotted the Livengood Highway, which we followed to Fairbanks. I landed at Weeks Field, although there was only a couple of inches of snow on the runway, which brought my ski-equipped T-Craft to an abrupt stop.

Jack found a ride to the hospital and left me on my own. I needed to refuel. Reluctant to taxi the plane because of the exposed gravel, I carried gasoline in five-gallon cans across the runway to the plane, requiring about an hour. I burned another hour shopping for beer and other items my friends at Bettles had requested.

With gas tanks full and shopping done, I filed a flight plan and took off, heading home. I estimated two hours of daylight remained, realizing I would be landing after sunset, but with snow on the ground and a clear sky, I thought I could find Bettles and land safely.

I hadn't counted on a 35-mile-an-hour headwind blowing in advance of a major snowstorm moving in from the northwest. This slowed me, and my flying time from Fairbanks to the Yukon River was nearly two hours. As I passed over that broad frozen river, the weak winter sun sank out of sight. I still had ninety miles to go.

It was soon too dark to read my compass. I had a flashlight with fresh batteries and used that for a time, but because of the extreme cold the batteries soon gave out.

Bettles didn't have an airways beacon at that time, and the few electric lights at the CAA station weren't enough for me to home on. I didn't know the country well, and without a compass and without landmarks or a light to guide me, I decided the safest place for old Andy was on the ground. I located the largest frozen lake in sight and felt my way down to a good landing, planning to continue my flight the following morning.

Sitting inside the silent, rapidly cooling airplane, I nibbled a few crackers from my emergency gear. Lacking a sleeping bag, I wrapped myself in my engine and wing covers. Fortunately I wore woolen long johns, a fine down-filled parka, and down-insulated pants, as well as arctic boots.

Despite the warm clothing, I spent a chilly night, exposed to temperatures of around 50-below zero.

I was thirsty, and although I had six cases of beer aboard, I didn't drink any of it, for I was and am a teetotaler. It was a silly situation, being thirsty with those cases of beer in the plane. Subsequently I learned that drinking alcohol causes blood vessels to dilate, with subsequent heat loss, so I actually did well to not drink that beer.

At daybreak I found ten inches of light dry snow on the plane's wings. It was still snowing, with two miles visibility. The plus side of that was the rise in temperature to 30 degrees below zero that came with the clouds and snow. I had no heater for the engine, and it was too cold for the engine to start; the oil had congealed. I decided to heat the engine with a small wood fire.

I didn't want to set the plane afire, so I examined the engine to be sure there were no gasoline leaks or accumulations of oil. Then with several large armloads of dry spruce branches broken from the bottoms of trees around the lake, I lit a small fire inside the confines of the

engine cover, which draped to the ground. I added a few sticks of wood at a time, guarding against burning the plane or the engine cover.

After several hours the engine was almost ready to start. Suddenly flames burst from the front of the cowling. I was sure I had lost my airplane. I would be stranded, perhaps forever.

Then, strangely, the flames disappeared.

The cause? In winterizing, Jim Magoffin had fitted a small

Snow-covered peaks of the high-country Brooks Range shine above Agiak Lake in August. The hunter is Dr. Edward Wiegand, of Sandusky, Ohio. I flew him and his guide to the lake. JIM REARDEN PHOTO

square of Plexiglas over one end of the air intake. It had overheated and ignited, then gone out. There was no damage.

The engine started on the second swing of the prop, and I took off without difficulty and followed a compass heading. But I had been overly optimistic about the visibility and ceiling. To keep the ground in sight, I had to fly just above the treetops. Visibility seemed to be diminishing. After flying about ten miles I realized I was pushing my luck, so I landed on another large lake.

I was hungry. In my survival groceries was a four-pound bag of pancake flour, but no skillet or grease. I built a small fire at the edge of the lake and melted snow in my only pan—a pie tin. This gave me water to drink and to mix with pancake flour.

After drinking all the water I wanted, I poured in a little pancake mix, heating and stirring. Not a gourmet meal, but the resulting chunks of cooked batter eased my hunger pangs.

My second night in the plane, at 30 below, was a bit more comfortable. I wasn't thirsty or hungry, and I shifted my load of beer and other items so I could stretch out better.

Daylight broke clear and cold. Again I carefully preheated the engine with a small wood fire. I was ready to prop the plane when a Civil Air Patrol plane en route to Bettles to launch a search for me passed directly overhead. It was flown by, of all pilots, Jim Magoffin. He landed, taxied close, and climbed out, leaving his engine running.

"Well, Andy, don't tell me the engine on the T-Craft I sold you has quit," he joked.

"Jim, the instrument lights don't work," I joked back. With no generator or battery, the plane of course had no instrument lights. "I couldn't read the compass when it got dark so I had to sit down."

With the joking over, he told me I was on course and only forty miles from Bettles. We took off, and I followed him the rest of the way.

Jack Irwin lost a couple of toes at the Fairbanks hospital and three weeks later returned to Bettles. The CAA docked me two days of pay for unauthorized leave. The flight taught me valuable lessons; I knew now that I must be better prepared if I wanted to continue flying in the Arctic.

I immediately added to the emergency gear in the T-Craft. For a time I was somewhat timid about venturing from Bettles on my flights of exploration. I continued to make charter flights, mostly hauling passengers between villages along the Koyukuk River where I could hardly become lost.

MARCH ROLLED AROUND, and with it, longer daylight hours and moderating temperatures. The land was covered with deep snow, making it possible to land on skis almost anywhere. I grew bolder and extended my range. Then word came that a migrating caribou herd was in the foothills of the Brooks Range north of Bettles. We were low on meat at the CAA station, so Canuck and I took the T-Craft to see if we could bag a caribou.

With plenty of emergency gear aboard, we left Bettles early and flew up the John River into country new to both of us. We saw no caribou, but continued flying north, enjoying ourselves on a beautiful sunshiny day. We tried to follow an aerial map, but had to guess where we were; in 1948 these maps were crude.

Eventually we flew through a pass in the Brooks Range and found ourselves in the northern foothills—now known as the North Slope, and famous for its fabulous oil production. We soon spotted a herd of caribou. I landed on hard, wind-packed snow, but by the time the airplane stopped sliding and we had rifles ready, the animals had fled.

I had been carefully watching the fuel gauge of the nose tank, which held twelve gallons.

"Canuck, we'd better head back. We're down to almost half our fuel," I said.

"OK," he responded. "I don't know where we'd put a caribou if we got one." He was referring to the pile of emergency gear I had stuffed into the plane. He was right. I'd have had to take Canuck home and return alone for a 150-pound caribou if we got one.

Weather was still calm and sunny when we took off, and we decided to take a direct heading for Bettles instead of following a winding pass. I climbed the T-Craft above the highest peaks of the rugged mountains. The scenery was spectacular, with peaks and valleys of that wild range extending east and west as far as we could see.

As we passed over the summit of the range, I descended into a south-trending river valley. We were uncertain of our location, but continued to follow the river south. At this point the gauge for the nose fuel tank was getting low.

I switched to the reserve gas tank in the right wing, which held six gallons. Gasoline was supposed to flow from it into the nose tank. But the nose tank gauge didn't move up; instead it continued to gradually drop. Was the wing tank empty? Had we unknowingly been burning fuel from it?

I decided to land while we still had fuel in the nose tank, picked a straight stretch of the river, and eased the plane down. We walked about stretching our legs and picked up the tail of the airplane to turn it in the direction I wanted to take off. As we did, we heard gasoline draining from the wing tank into the nose tank. That resolved our problem; a bit of ice or dirt must have plugged the line.

In the air again we started seeing dogsled trails and guessed we had followed the Alatna River out of the mountains, putting us near

Allakaket, a missionary village about forty miles southwest of Bettles on the Koyukuk River. Sure enough we soon passed over Allakaket and we followed the Koyukuk River upstream and home to Bettles.

That first flight into the Brooks Range left me with a vivid impression of these spectacular mountains that arc across northern Alaska. At the time probably fewer than five hundred people, mostly Eskimos, lived in this California-size region of peaks and valleys. Another several thousand people, mostly Indians and Eskimos, lived along the southern lowland rivers that drained the mountains, as well as on the northern Arctic Ocean coast. Most lived from the land, utilizing game, fish, berries, and sea mammals for sustenance. It was— and still is— the last great wilderness frontier of the United States.

That one flight convinced me perhaps more than any other single factor that I was in the right place at the right time. The opportunity was there for me to fly commercially, to provide a needed service for people who were already there and for those who were bound to come to this spectacular region. Clearly this was a perfect place for airplanes; ground travel in this vast land could never compete with swift and efficient air travel.

I flew my T-Craft for about a year, but for commercial flying of passengers and freight I needed an airplane with more room and greater weight-carrying capability. I sold my plane to Ed Klopp, station manager for the CAA at Bettles. By then Wien Airlines had provided me with a plane more suitable for commercial work.

WIEN AIRLINES

My decision to resign from the CAA so I could fly for Wien Airlines left me out in the cold, literally, for I had to move out of government housing. I expected my proposed flying business for Wien to develop at Bettles Field, five miles from Bettles Village. There was no housing at the field, so I bought two tents. That summer I lived in one and situated my "office" in the other.

Wien Airlines pioneered flying in Alaska. The beginnings of the company extended back among the very first airplane flights in the Territory. Commercial flying started in the 1920s when Alaska was a nearly roadless land with widely scattered settlements, gold and copper mines, trading posts, and traplines.

Noel Wien arrived in Alaska in 1924 and made the first flight ever between Anchorage and Fairbanks, in an open-cockpit Hisso-

It was time for another mail flight as I walked out the door of my Bettles Field roadhouse

on this day in 1955. The logs for this building were sawed on three sides and pinned

together with three-quarter-inch pipe. Caulking had to be constantly renewed because

the logs expanded and contracted with the seasonal temperature change that could

amount to 150 degrees. JIM REARDEN PHOTO

powered Standard J-1 biplane. Once in Fairbanks, Noel made many flights in the Standard to outlying camps, villages and traplines. At the time, most travel in the Territory was by paddle-wheel riverboats in summer and dog and horse teams in winter. A 400-mile wagon road linked Valdez to Fairbanks. The 470-mile Alaska Railroad from the port of Seward was the only link between Fairbanks and Anchorage.

In 1927 Noel and his brother Ralph, along with Gene Miller, formed Wien Brothers and started air service between Nome and Fairbanks with an open-cockpit Standard biplane that cruised at 60 miles per hour. Traders, miners, salesmen, missionaries, reindeer herders, and fur buyers were among the first passengers.

In 1928 Wien bought Hubert Wilkins' Stinson cabin biplane powered with a Wright air-cooled engine. Wilkins had stored it in Fairbanks after an unsuccessful attempt to fly from Alaska to Norway. With this plane Wien made weekly round trips between Nome and Fairbanks. One-way fare was three hundred dollars.

That fall Noel hauled a thousand pounds in passengers and freight in one load across the Arctic Circle to the villages of Bettles and Wiseman, the largest commercial load ever to leave Fairbanks in an airplane. Wien soon had contracts to haul mail from Fairbanks to Nome—the first airmail contract issued to an Alaskan commercial airline.

By February 1929, Ralph Wien was flying charters out of Fairbanks with a four-place Stinson biplane, while Noel was making the weekly round trip between Fairbanks and Nome, chartering his plane at both ends. About then the company purchased an all-metal eight-place Hamilton plane.

With this plane, Noel Wien made a daring flight from Nome to East Cape Siberia in February 1929 to rescue a load of furs from a trading ship locked in the ice. It was the first round-trip flight between North America and Asia.

In December 1930, Noel was accompanied by his younger brother Sigurd as they flew a new Stinson Jr. from Minnesota to Fairbanks. Noel had taught Sigurd to fly in 1923, and this was Sigurd's first trip to Alaska.

On August 17 and 18, 1935, flying a Bellanca, Noel and copilot Victor Ross flew Nome entrepreneur and photographer Alfred Lomen on the first commercial passenger flight between Fairbanks and Seattle. Lomen carried the last photos taken of Will Rogers and Wiley Post, as well as pictures taken at the scene of the tragic crash near Barrow that killed these famous men.

Sigurd Wien made his first commercial flight in 1937 when he flew a 1936 Cessna the 250 miles from Fairbanks to Wild Lake, in the Koyukuk country. He was soon located at Nome, where he provided aerial service to people on the Seward Peninsula and from the Bristol Bay area as far north as Barrow.

Sig got along well with the Eskimos. His thoughts ran in a direct line like theirs. He was short on words, but long on performance. At least one Eskimo baby was named Sigwien by parents who idolized Sig. It was Sig who, in the early 1940s, first started to fly supplies to the Anaktuvuk Pass Eskimos. At the time only one of these Native wanderers spoke English.

In 1945, two years before I arrived at Bettles, Wien Airlines acquired two Douglas DC-3 transport planes. The company continued to expand as the major airline serving Interior and Arctic Alaska during the 1950s and 1960s.

Over the years the company went through many business transformations and was called variously Wien Brothers, Wien Alaska Airways, Wien Airways of Alaska, Wien Air Alaska, Wien Alaska Airlines, Wien Consolidated Airlines, and Wien Airlines. Noel, Sigurd, Ralph, and Fritz were the four Minnesota farm boys who

This is how the roadhouse I built and operated at Bettles Field looked in the mid-1950s. It took three years to thaw the permanently frozen ground for the basement and to build the 40-by-60-foot log building. The roadhouse served as a combination hotel, restaurant, airline headquarters, and post office, and it was key to Bettles becoming a major business hub in the Koyukuk Valley. It's still operating today.

NOEL WIEN PHOTO

sparked the airline; another brother, Harold, worked briefly for the company, then returned to farming in Minnesota. Ralph Wien was killed in an airplane accident at Kotzebue in 1930.

In 1940 Sig Wien became president and general manager and he held those positions for thirty years. He oversaw the airline's development from a few primitive single-engine bush planes into the jet age, which came in 1968 when the company bought two Boeing 737 twin-jet airplanes.

UPON BECOMING A WIEN PILOT, I was called to headquarters at Fairbanks. After learning something of policy and procedures, I was sent to retrieve a Cessna 140 single-engine, two-place airplane the company owned that had been left at Fort Nelson, British Columbia.

When I got to Fort Nelson, weather was still bad, and I had to wait a couple of days before I could take off. Several other small planes that had been delayed by the weather left with me about 4 A.M. on a bright, clear morning. We hoped to reach Fairbanks that day.

Canadian regulations require that small planes follow the Alaska Highway, where it will be easier to find any plane that goes down. I carried several extra five-gallon cans of fuel in the cabin, and I stopped at airfields along the route to refuel.

Whitehorse was my last stop in Canada. I refueled, cleared Canadian customs, and was ready to make the last long hop to Fairbanks. Then the pilot of one of the other planes suggested we go into Whitehorse for lunch. It was well past noon, and food sounded good.

I caught a cab to the restaurant and was enjoying lunch when a customs agent came to the table and said, "The chief customs agent wants to talk with you when you're through there. His office is upstairs."

I finished my meal and sauntered upstairs to the man's office. "I'm Jim Anderson. You wanted to see me?"

"You stole that airplane at Fort Nelson, didn't you?" he blasted at me.

I truly thought he was kidding.

"Why, yes, I suppose I did," I kidded back.

He wasn't kidding. "OK, wise guy, prove to me that you have a right to that airplane."

I had left the company's letter authorizing me to take the airplane with the commanding officer at the air base at Fort Nelson.

"I need identification," the customs man said.

I had left my pilot's licenses and other credentials in my T-Craft, parked at Weeks Field, Fairbanks.

I explained my problem.

"We're going to hold you and the airplane until we see what's going on," he announced.

In the end I paid for a telephone call to the commanding officer at Fort Nelson and to Wien at Fairbanks. The customs officer did all the talking. He released me and the plane. Then he said, "Never show your face in Canada again without proper identification." I delivered the plane to Fairbanks later that day.

I returned to Bettles, where I was soon providing transportation for villagers, miners, and others for hundreds of miles around. I kept in touch with my customers by radio, for most villages and some of the miners in the region had two-way radios.

Fall neared, and there was a distinct chill in the mornings when I crawled out of my cot in the tent. In Bettles, at a latitude of 66 degrees, 54 minutes north, snow can fly in August, at least in the higher elevations. Living in that tent, I began to wonder what I was going to do when winter arrived.

Then Sig Wien flew to Bettles to look over my operation.

"I've got to have a building to live in soon if I'm going to spend the winter here," I said.

"Sorry, Andy," he said, "We don't have money to put up a building right now. We're struggling, and trying to expand into new routes and buy new planes."

I had saved a few dollars and suggested to Sig that I advance the company $2,500 for material to construct a building. It could serve as temporary housing through the winter, and later it could become a freight shed for Wien.

"Fine, Andy. It's a deal. We'll repay you when we take over the building. I'll order lumber when I get back to Fairbanks, and we'll fly it to you."

Canuck agreed to help construct the building in exchange for two round-trip airline tickets to Fairbanks. I thought I was all set for winter as we waited. But weeks went by and no building materials arrived.

In time I learned that Sig had gone south on vacation and somehow had neglected to order my building materials. I had no more funds, and the struggling Wien company had no credit with Fairbanks suppliers.

In my life I've had many interventions of what might be termed "help from above." Some would call it luck. In this case, intervention came from a project engineer for the CAA, which was building a new weather station and housing at Bettles Field. Building materials had been flown to the field for the new station, but winter was near and the CAA construction season had ended. My angel—Bob Matson— visited me in my tent one evening.

"Andy, I realize your problem. I'm sticking my neck out a mile and putting my job in jeopardy. But if you'll promise to replace everything no later than early spring, I'll loan you enough building supplies from material I have on hand for you to put up your building."

It truly was a gift from heaven.

"If I have to carry that lumber from Fairbanks to Bettles Field on my back, it'll be here next spring," I vowed. The lumber was duly replaced before spring.

The shed we hastily erected in early winter of 1948-49 was the beginning of the Wien station at Bettles. I moved into it in late November when nighttime temperatures were well below zero. It was insulated and easy to heat, albeit primitive and lacking in space. I dug a cellar under it and installed a well. To accomplish this I had to thaw the frozen ground with a steam boiler. Since the shed sat on skids, it proved impossible to seal cold from the "basement," and the well froze during the first winter.

For two winters I lived in that shed, melted snow for water, used an outhouse, bathed in an old-fashioned round laundry tub, and heated with an oil-burning cook stove. I didn't complain—it was better than the tent.

Winter flying was a time hog. All I had for heating the plane was a plumber's gasoline-fired stove, called a "pot." I didn't dare leave it burning unguarded for fear the plane would catch fire. I had to get up at 4 A. M. and heat the plane's engine for four hours before I could fly.

I waited months for Wien to reimburse me for the money I had advanced. When it didn't materialize, I agreed to accept Wien Airlines stock in lieu of cash. I was also reimbursed with Wien stock for other items I advanced money for—fuel, more building materials, and so forth. This arrangement continued for the entire time I flew for Wien.

THE FLOW OF TRAFFIC WAS PHENOMENAL from the very beginning. And with the large number of passengers that Wien flew to Bettles Field, mostly bound for villages up and down the Koyukuk River, it became obvious a roadhouse or a lodge was needed for feeding and housing overnight guests. I decided to build one on CAA land next to the airport runway and, on my own, leased the land.

Much of Alaska's arctic ground is permanently frozen, called permafrost. To dig six feet down for a basement, I used a bulldozer to scrape dirt from the site as it slowly thawed. This required three summers. At six feet I hit gravel, which proved to be a good foundation for the lodge footer logs.

Wilford Evans, a resident of the nearby village of Allakaket, owned a portable sawmill. For $5,000 he agreed to construct the 40-by-60-foot building from logs slabbed on three sides. This included windows, doors, subflooring, and roof. Completing the interior of the building, with fixtures, ran the cost to $17,500.

Over a period of three years I built my log roadhouse, or lodge, which had two main floors and a basement. Spruce logs, cut within thirty miles of Bettles, formed the walls. They were pinned in place with three-quarter-inch pipe. Upstairs were two bathrooms and six bedrooms with a large lounge.

For myself, the station manager, I had a two-bedroom apartment with living room and bath on the first floor. Most airline business took place in a large lobby/waiting room and office. Nearby was a large table where we served family-style meals. A small gift shop, post office, and a basement laundry completed the layout.

Eventually Wien Airlines bought the roadhouse and leased it back to me. In lieu of cash I accepted stock in the company.

Later, business increased and during summer we erected a 16-by-20-foot tent with a floor, walls, and heat that served as a second dining room.

Winter temperatures often plunge to 65 and 70 degrees below zero at Bettles, so I needed a good source of heat. I had two heavy-duty fifty-gallon oil drums welded end to end, with a fire door opening at one end and a smokestack fitting on the other. It stood on four stout metal legs. I built a sheet metal heat jacket around it, and ran large ducts to various parts of the lodge.

We burned at least forty cords of wood a year in that huge stove. A spruce-birch forest extends for hundreds of miles in all directions from Bettles. I paid local residents forty dollars a cord to cut my firewood, which was free for the taking. Trees were felled, trimmed, and stacked in summer. In early fall after freezeup, but before heavy snowfall, a tractor hauled the stacks of trees to the lodge, where they were cut into four-foot lengths to fit our oil drum stove.

The Navy Seabees who built the runway drilled to ninety feet without finding water. Old-timers, mostly miners, told me there must be water at my lodge site, for the Koyukuk River was only a quarter of a mile away.

I decided to dig a well in the basement, knocking the ends out of fifty-gallon drums and using these as casings as I dug. The gravel was damp at sixteen feet. I ordered a well point designed to be driven through gravel and drove it to about thirty feet. With a string and weight I measured about eight feet of water in the pipe. But the pump I used couldn't pull the water out.

Puzzled, I pulled the well point and discovered the screens were clogged. I had apparently driven through silt or sand. I cleaned the screens and again drove the point, measuring water depth as I went. This time I frequently used a hand pitcher pump, which removed the sand. Then I would drive the point another two feet. It took two weeks to drive ten feet, but I ended with a beautiful well that produced unlimited, icy, pure water. With the well in the basement, all water lines were inside, hence I had no frozen pipes.

Sewer lines were six feet deep and encased within a larger pipe. The larger pipe ran from the inside of the basement to the septic tank. In event of a frozen sewer line (which we never experienced), we could have pulled the sewer pipe back into the basement to thaw. Once the lodge was in full operation we had to pump out the log septic tank about every ten days—a small price to pay for the luxury of indoor plumbing.

One autumn while the lodge was still under construction, my mother visited me. We were using an outdoor privy. "James, what do you do in the winter when you have to visit the outhouse?" she asked.

"Mom, the first thing is to make certain you have to go. That cuts down on exposure time."

She just shook her head. She couldn't imagine visiting an unheated outdoor privy at 50 and 60 degrees below. I, on the other hand, figured that to be one of the minor challenges of frontier life in arctic Alaska.

When the CAA moved to Bettles Field from Bettles Village in 1952, they installed a large generator from which we bought electrical power for the roadhouse.

Almost as soon as we had completed the lodge, a construction gang arrived to build more CAA facilities at Bettles Field. That summer we served an average of 140 meals a day for three months straight. Within a short time the Air Force established a navigation station at Bettles Field with eight technicians to run it. We fed and housed these men, a nice boost in business.

I charged fourteen dollars a day for three meals and sleeping quarters. I hired a cook and a maid. When business increased I hired a third person.

The roadhouse by itself wasn't especially profitable, but having it made it possible for me to fly a steady stream of mail, freight, and passengers to the many villages and mines in the region. Partially because of the roadhouse, Bettles developed into an important rural Alaska transportation hub.

BUILDING A BUSINESS

I wasn't so busy flying in the Koyukuk Valley as a Wien pilot that, over several months, I couldn't court and win the hand of a lovely Eskimo girl named Hannah Tobuk, who arrived at Bettles to visit relatives. Hannah had been educated in Seattle and had lived there for some years, but Alaska was always her home. Hannah made the perfect partner for my life at Bettles. We had a comfortable apartment in the roadhouse, and Hannah saw to it that roadhouse life went smoothly for our growing number of guests.

We became the parents of two boys and two girls. Sadly, our second girl, Patricia, died at the age of fourteen after living most of her years in a personal care facility. The other children, Mary, Philip, and David, spent their early years growing up in the roadhouse, spoiled and loved by us as well as by our many guests.

My friend Canuck Killen was hired by Wien and remained with us at Bettles for many years. He was in his mid-thirties when I met him, and he could do just about anything. He was a gifted carpenter, machinist, mechanic, and all-around handyman. Sometimes he even

At Bettles, I refuel a float-equipped Cessna 180. JIM REARDEN PHOTO

The Anderson family gathers for a photo at Bettles in 1955. Mary is on my lap, while Hannah holds baby Phil. JIM REARDEN PHOTO

entertained our guests with his renditions of the poems of Robert Service. In later years Canuck went to aircraft mechanic's school in Los Angeles and he became one of Wien's most respected aviation mechanics.

Canuck and Jack Irwin had built a cabin at Bettles Village and were nicely settled in for winter when I arrived in late 1947. I enjoyed visiting with them and hearing their tales of life in Fairbanks. I especially enjoyed Jack's story about serving a hotcake breakfast to friends. He was cleaning up afterward and noticed his stirring spoon standing straight up in leftover pancake batter. He pulled, and the spoon with attached batter came free of the pan. Startled, he studied the situation and realized to his horror that he had two similar bags—one containing pancake flour, the other plaster of Paris.

No one got sick, and Jack was afraid to tell his friends of his mistake. Joking, he told me, "That was one reason I left Fairbanks. I had a lot of very firm friends and I decided I'd be better off getting away from them."

Nellie, Canuck's wife, enjoyed the northland as much as he did. Their enjoyment included a month of hunting each fall. It was a ritual we expected each September so we made plans accordingly, knowing the Killens would be gone until they had their game or wore themselves out from the trip.

With their riverboat they headed up the John River where Dall sheep and moose were abundant. While Canuck hunted, Nellie picked gallons of blueberries and cranberries (lingonberries), which they preserved for winter use.

I was surprised one September to see Canuck back at Bettles Field a few days after he and Nellie had left for their annual hunt. He seemed nervous.

He told me that from the John River he had spotted a band of Dall sheep on a nearby high ridge. He left Nellie on the opposite side of the river where she could pick berries, crossed the river, tied his boat, and climbed the steep slope to the sheep. He was carrying a newly purchased rifle.

Near the mountaintop, Canuck dropped a nice ram with one shot. He tried to eject the shell casing, but it was jammed. He spent some time trying to remove the case, then gave up. He dressed the white sheep and started dragging it down the mountain with a harness.

Near the bottom of the mountain the carcass hung up occasionally on trees and bushes. Each time, Canuck yanked it loose without difficulty. But at one point, the carcass refused to budge despite Canuck's strong yanks. He turned around and saw the sheep held in the jaws of a huge grizzly bear.

His rifle jammed and useless, Canuck swatted at the bear with the rifle butt. The bear swatted back with a ham-size, white-clawed paw.

Canuck dodged. Since he was tied to the sheep, his movement yanked the sheep. The bear, thinking the sheep was getting away, forgot momentarily about Canuck and again grabbed it.

With his hunting knife, Canuck cut loose from the harness and sprinted toward his boat a few hundred yards away. He peered over his shoulder several times and each time saw the bear close on his heels. He arrived at the boat, cut it loose, gave it a great shove and leaped aboard, leaving the bear growling on the beach.

Shakily he crossed the river and picked up Nellie. They returned to Bettles Field to spend their vacation in more peaceful surroundings.

Another long-time trusted employee was Hannah's uncle, Frank Tobuk, a tall, handsome Eskimo who refueled and loaded our aircraft. He also kept the basement full of firewood for the furnace, and he kept the lodge weather-tight by frequently chinking the logs, which shifted under the impact of seasonal temperature changes of around 150 degrees. Frank was knowledgeable, dependable, and scrupulously honest. He didn't talk a lot, but when he said something it was worth listening to.

SOON AFTER I STARTED FLYING FOR WIEN, the weekly DC-3 flights to Barrow, with a stop at Bettles, increased to three a week to handle an ever-growing volume of freight and passengers. Sometimes more freight or passengers stopped with us than were bound for the large Eskimo village of Barrow at the end of the line.

With the roadhouse completed and operating, the freight shed I had lived in now functioned as intended, except that we built living accommodations in it for Wien mechanics who often flew from Fairbanks to work on our aircraft.

My daughter Mary and son Phil smile in front of the sign at Bettles Field. My wife, Hannah, and I used the photo as a Christmas card from the Anderson family in the mid-1950s. JIM REARDEN PHOTO

In addition to new contracts to fly U.S. mail, gold mining boomed, which resulted in many passengers and much freight, especially in summer. Clearly I had blundered into a golden opportunity to provide a much-needed service.

I bought a farm tractor with a hydraulic high-lift attachment—for which Wien credited me with stock in the company—for off-loading heavy freight from the mainliner DC-3s. It was followed by a military-type Jeep that I used to haul supplies and passengers to the floatplanes that I kept on the Koyukuk River a mile or so from the roadhouse. There was no need for an automobile, for I flew everywhere; I didn't own a car for twenty years.

We had an informal family relationship with guests at the roadhouse. Miners, hunters, and Native residents of villages we served felt at home there. Often they helped load or unload the DC-3 flights. In winter it wasn't uncommon for a guest to dive into the basement to feed the ravenous woodstove. Others, unbidden, sometimes helped in the kitchen where we normally employed a cook and a helper.

In the busy season we often worked around the clock. We usually kept three types of aircraft at the station; in the early years it was often a Cessna 170, a Cessna 180, a DeHavilland Beaver, and the Wasp-powered freight-carrying Noorduyn Norseman. I seldom flew less than 150 hours a month; summers I averaged 250 hours a month in the air.

Sometimes delivering freight that came in the DC-3 mainliners was a problem. Harry Leonard, postmaster at Wiseman, ordered a Sears Roebuck farm wagon that wouldn't fit inside our biggest plane, the Norseman. Finally we tied it to the Norseman's landing gear. I managed to ease the Norseman into flight, but the blunt wagon was a terrible wind-catcher. Though I used most of its 600 horses, the big plane flew just above stalling speed, and the thirty-minute flight to Wiseman stretched into more than an hour. After that we often tied too-large-to-fit-inside freight items to the outside of our planes for delivery.

Arctic flying in the 1950s and 1960s was challenging and I could easily have developed a "worrying mind," as the old song goes. Poor flying weather was my single greatest problem. But the pace of life for residents in arctic Alaska is mostly relaxed and slow. I learned to be patient. My customers understood as well as I did that weather dictated what we did. I learned not to fret when bad weather kept me grounded; it would improve, and when it did, I could go on with my flying.

For many years I was the only bush pilot providing consistent service to the people in the upper Koyukuk Valley, and the only pilot who could be reached regularly by radio by villagers, miners, and others in the region. Before I arrived with my swift airplanes, most local residents traveled on foot, by dog team, or by riverboat. Instead of hours and minutes of travel time between villages, mines, or camping and hunting areas, weeks and even months had been required.

I OFTEN MADE MEDICAL EVACUATION FLIGHTS at night or during terrible weather. Sometimes both. Why, I never understood, but medivac flights seemed more frequent in nighttime or during the worst possible flying conditions. Weather stations and navigation aids were spaced so far apart they were almost useless to me. As a result I had to learn the area, including the location of every mountain ridge, every lake, and every possible landing site.

One midwinter night a radioed request came from Allakaket for me to fly a young woman in labor and expecting her first child to the Alaska Native Service hospital at Tanana. It was snowing, with a blustery wind, and the temperature was well below zero—a nasty night; even a daylight flight would have been questionable.

There were no lights at the short dirt runway at Allakaket. I had to depend on a Good Samaritan to stand at the approach or touchdown end of the runway with a flashlight pointed in the direction I was to land. This allowed me to fly so the landing light on the aircraft could guide me in landing—light from the flashlight helped me to determine how close I was to the ground. On a long airfield there would have been some leeway, for then I could have felt my way down, gradually allowing the plane to descend. But long runways were an almost nonexistent luxury in the Koyukuk. The hummocky strip at Allakaket was no exception. Even today the airport facility directory warns pilots landing at Allakaket to "Watch for children and dogs on runway" and that "river floods south end in spring."

On the forty-mile flight to Allakaket I flew through the snow and wind a few hundred feet above the white, frozen Koyukuk River, bordered by dark trees. Someone, I never knew who, was waiting in the dark with a flashlight to show me where the end of the runway was. I let down and managed to stop. Several villagers were there with the frightened young mother-to-be wrapped in furs and blankets.

I didn't get out of the plane or stop the engine. Someone helped the young woman in and I pulled the lap belt across her upper legs and snapped it in place. I tried to reassure her with a smile, swung the plane around, and roared down the runway with the snow flying every which way.

That 150-mile night flight from Allakaket to Tanana was one of the most hazardous I can remember. There are no predominant landmarks along the route, which consists of rolling hills interspersed with wooded areas. Here and there is an occasional treeless, windblown opening.

On my right was 3,500-feet-high Utopia Creek Mountain. Peaks on my left were four to five thousand feet high. To be sure I was above all of these I started climbing. I was almost instantly flying on instruments, with no sight of the ground. I climbed above five thousand feet and took up a compass course, praying there was no ice in the clouds where I flew.

The radio range at Tanana was not very reliable from fifty miles out, and my low-frequency receiver left a lot to be desired. For me to receive any signal at all I had to reel out forty feet of trailing antenna.

As we flew in the clouds and snow, I continually inspected the wings with a flashlight, praying ice wouldn't develop, and it didn't. I had no idea of wind direction and velocity at my altitude. Turbulence bounced the plane like a sock in a washing machine. My entire being was concentrated on flying and on the radio range signal. It gradually became stronger, indicating I was nearing Tanana.

After flying as if inside a bottle of milk for more than an hour, and with the radio range signal coming in loud and strong, I called CAA flight service and requested the runway lights be turned to high intensity. I caught a glimpse of a glow below and cautiously descended in a tight circle, keeping the glow in view. I broke out of the clouds a few hundred feet directly over the runway. Landing was no problem.

A Jeep arrived to pick up my passenger, who had not said a word. I assumed she was justifiably frightened half out of her wits. She did manage a wan, "Thank you, Andy," as she left. She delivered a healthy baby boy just ten minutes after arriving at the hospital.

"You have to leave immediately, Andy," one of the attendants said. "The community is under a medical quarantine."

I figured I had used up all my luck that night and I wasn't about to try to fly the 190 miles back to Bettles in the snowy darkness. It was too cold to sleep in my plane, so I persuaded the man on watch to isolate me in the CAA building, where I grabbed a few hours of sleep. At daylight I was able to return to Bettles.

I flew so many women in labor to the Tanana hospital that the doctors there gave me instructions on how to handle a birth and provided me with an emergency kit to keep in the plane. Happily I never had to use it, for no baby was ever born on any of my flights. However, birth was too near to suit me on a number of occasions.

I finally announced to everyone in the region that I would no longer consider a woman in labor a medical emergency. This resulted in my flying expectant mothers to the hospital in advance of expected birth dates, and these flights were made in decent weather and during daylight.

WHEN I FLEW ON A COOL, CRISP autumn day, with the land below me in splotches of vivid orange, red, and gold, with the air as smooth as silk, the joy within me seemed to overflow. In the crystal air of the Arctic I could sometimes see for a hundred miles or more. The beauty of the land under my wings always gave me pleasure.

The Brooks Range of northern Alaska was, and still is, the wildest, most remote, and least-hunted of the important big-game areas in North America. This rugged, untouched wilderness, the northern extension of the Rockies, spans six hundred miles across northern Alaska and has a north-south depth of about one hundred miles.

During the 1950s and 1960s, increasing numbers of adventuresome hunters arrived at Bettles, asking me to fly them into this lonely land. Vast herds of roaming caribou are found there; I've seen caribou crowded so closely together I would have had difficulty in finding an open spot among them for landing an airplane. At times on clear fall days I could see the snowy white necks of the bulls from a distance of twenty miles.

Grizzly bears also wander along the mountains, mostly at or beyond timberline. Some are blond, some dark. Others are blond on the back, with dark legs. Shiny black bears also roam the south slope of the range.

Pure white Dall sheep—North America's most northerly wild mountain sheep—dot the high ridges. Early in fall on grassy peaks, they look like popcorn scattered on green cloth. After snowfall, the only way I could see them was to find their tracks; their bodies are nearly invisible against snow.

Moose are most abundant on the south slopes of the mountains, although these huge deer are also found on the North Slope. As I

*A Noorduyn Norseman sits on its floats on the bank of the Koyukuk River at Bettles as
Warren "Canuck" Killen refuels the plane.* ANDY ANDERSON PHOTO

flew above them, their salt-and-pepper and dark bodies stood out
against the spruce, birch, and open tundra. Often the bright, clean
antlers of the bulls flashed like mirrors in the distance. Sometimes I
looked down in wonder at these big, ugly-handsome deer as they
stood with heads submerged in lakes, feeding on aquatic plants.

Wolves were especially abundant in the Koyukuk Valley, for they
are found everywhere there are moose or caribou, their main food.

Warren Tilman, an old-time Alaska guide and one of the few
who hunted these mountains in the 1930s, once said, "In the Brooks
Range the game dies of old age and the country remains just like God
put it there."

Wien Airlines and I started to promote hunting in the region.
Within a few years as many as sixty to seventy hunters and guides

were chartering my float-equipped airplane to reach isolated spots of the mountains each fall. Hunting season ran from the first of August until late September. Whenever the nonresident hunters or guides didn't want the meat of big game they killed, I flew it to the nearest Native village, where it was welcome indeed; none was wasted.

Some hunters arrived at Bettles unprepared for camping in mountains where snow can fly every month of the year, and I often loaned them the proper equipment. Timberline in the Brooks Range is at about 2,000 feet, and much of the range is above and beyond the last timber. There are twenty-four hours of daylight in the range during early August, but even then the temperature can drop fast; August weather is comparable to that of early October in the northern contiguous states.

High-country lakes often freeze over in early September, halting floatplane landings. After a few years I ruled that my hunters had to be out of the high-country Brooks Range by September 1. If they weren't, I told them, I refused to be responsible if they became frozen in. In this event, their only recourse would be to walk out, and a hundred-mile hike through snow-covered mountains could be deadly. (My annual changeover to skis on my airplanes usually came in October or November—too late to benefit any hunters who failed to beat the freeze.) Happily, no hunter I flew into the mountains ever had to hike out—thanks, I believe, to my September 1 deadline.

Some of Alaska's top guides came to Bettles Field with their hunters. I recall providing charter service to Hal Waugh, Warren Tilman, Jim Rearden, and the famed partners Bill Pinnell and Morris Talifson, and there were many others.

My most famous passenger was General James Doolittle. He had been a hero to me ever since his famous bombing raid on Tokyo early in World War II. One fall he was a guest at an Air Force hunting camp not far from Chandler Lake. The Air Force had its own registered guides and floatplanes there, so I remained clear of them except for an occasional drop-in visit to check on their well-being.

On one such visit I learned that Doolittle was the only hunter in camp who had not taken a trophy Dall ram. I offered to fly him and his guide to nearby Shebou Lake, where I thought he would have no difficulty finding a trophy. I named this small, high-country lake Shebou, a combination of sheep and caribou, because of the constant abundance of trophy rams and caribou bulls.

The general and his guide climbed into my plane, and after a short flight I eased down for a landing on Shebou. This was long before Alaska's current regulation that prohibits hunters from hunting on the same day they are airborne. The two men headed up a nearby ridge while I waited. Later that day the ecstatic general, who was then in his 70s, and his guide appeared with the general's trophy ram, which we loaded into the plane and flew back to the Air Force camp.

Doolittle tried to pay me for the flight. This I refused, because my service was but a token of my appreciation for his wartime heroism. I did ask for and receive from him a one-dollar bill with his signature on it, which I treasure to this day.

THE WATERS OF THE BROOKS RANGE are a strong attraction to sport fishermen, drawn to the thousands of miles of clear-water streams and hundreds of untouched lakes.

Grayling abound in the clear rivers and most of the lakes. This handsome, flag-finned, rainbow-hued fish is usually eager to take a fly, a spinner, spoon, or small plug. White-fleshed and mild-flavored, it is a favorite of Alaska's sportfishermen.

Northern pike are abundant in most lakes and streams on the south slope of the Brooks Range.

JIM REARDEN PHOTO

Lake trout are plentiful at virtually every lake inlet and outlet across the range. They also cruise the shallows in summer. Some weigh up to twenty pounds, and they too make for fine eating.

Arctic char are found in most of the high-country lakes. This beautiful cold-water fish, related to the Dolly Varden, will strike at almost any lure. That toothy predator the pike, some up to thirty pounds, is abundant mostly in lakes along the south slope of the Brooks Range.

Sheefish, a close relative to the salmon, winter in coastal estuaries. They migrate annually to freshwater spawning grounds far up the clear Koyukuk River. Unlike the Pacific salmon, sheefish do not die after spawning. I often anchored a boat just above a riffle on the Koyukuk River and cast my lures just below the riffle. When sheefish were there, I would seldom cast without getting a strike. In the course of an afternoon I have caught upwards of four hundred pounds of these delicious, white-fleshed fighters. I enjoy sportfishing, and sheefish are among the best of fighters. But when I caught these fish it was to acquire a supply of them for winter use. No fish is tastier.

Whenever I returned to the roadhouse with sheefish, I removed the entrails, cut off the heads, removed the scales, wrapped the fish in aluminum foil, and dropped them into my freezer. They kept perfectly, and we could eat what passed for fresh fish all winter.

Koyukuk River salmon are mostly chums, or dog salmon, not normally considered a sport fish, for they rarely hit a lure. At spawning time when both grizzly and black bears flock to the tributaries to feed on these oily fish, I often flew low over these streams, enjoying the sight of these grand animals dining on their favorite food.

During my later years in Bettles, tourists occasionally arrived to charter my plane for a scenic flight. Those who accompanied me on such flights usually returned with stars in their eyes after viewing the splendors of the upper Koyukuk River country.

In all my work at Bettles, my principal obligation was clear: to offer transportation in return for a fee. When a customer called for service, he didn't want a weather report or a reason why I couldn't fly him or his freight. Nor was that customer concerned about the condition of my airplanes. He usually wanted one thing only: for me to fly him or his freight from point A to point B as quickly and safely as possible.

That was my primary goal for the seventeen years I flew from Bettles. However, in time I learned there was a bit more to it than that. My life became intertwined with that of my customers. Many became my close friends. Their triumphs and tragedies became my triumphs, my tragedies. On many flights, money was the last thing on my mind.

PEOPLE STRANGE AND BEWILDERING

On a scheduled mail run one winter, I landed at Chandalar Lake high in the Brooks Range. While waiting for the postmaster, I saw a couple of large sled dogs trotting across the frozen lake, pulling a sled that resembled an undersize television dish.

As they drew closer I made out a person sitting in the dish. He was heavily bundled in clothing because of the 40-below temperature. Frost on his parka ruff framed his face in white.

The dogs trotted to my plane and stopped. The man leaped to his feet, stuck out his hand, and said, "I'm Amero. I guess you're Andy."

His full name was Alfred W. Amero (pronounced AM-uh-row), but most people called him Old Amero. He was a Klondike goldrusher in 1898, making the journey on money from his mother. He didn't get rich in the Klondike, and he arrived in Alaska on the Kobuk River Stampede, a gold rush in 1898-99.

I flew my favorite bush plane, this float-equipped Cessna 180, to huge

Chandalar Lake in the high-country Brooks Range. The year is 1955.

JIM REARDEN PHOTO

Amero mined on the Kobuk River, the adjacent Noatak, and then the Koyukuk—and by the time I met him he had ranged all over the north. He was reputed to have been the best woodcutter in Beaver, where he had cut wood for the steamboats that once plied the Yukon River. He could take two pack dogs and go for weeks, even months, wandering in the mountains, searching for gold. Another old-timer once said Amero would size up a fat ground squirrel like it was a side of beef.

In 1946 Amero sold one of his claims for enough money to visit his old home in New England, the first time he had been south of the Yukon River since 1898. In all that time he had never been to Fairbanks, the only city within 250 miles. In his extreme old age he lived in the Pioneers' Home at Sitka.

We became friends. Although he was in his 80s, he moved, talked, and behaved like a much younger man. I even started calling him Amorous Amero, because he still had an eye for the ladies.

On one of my flights to Chandalar, he arranged for me to fly him to the Arctic Village area, about a hundred miles to the east. As a youngster he had worked for a mining company there, and the outfit had done well. Amero thought it might be worth reworking the claim.

He remained in the Arctic Village area for more than a month. When he returned to Chandalar he made no mention of his experiences at the old mine. Curious, I asked Red Adney, also of Chandalar, if Amero had found anything in the old diggings.

"Naw. All he found was a bunch of old horse manure," Red said.

The company Amero had worked for had used horses in their mining, and all Amero had found was what the horses left behind. He was proud of his ability to find gold, and his find at the old mine embarrassed him. He wasn't about to tell me about it.

ANOTHER CHANDALAR MINER I made friends with was Ellis Anderson. In his 80s when I first met him, he had arrived from Sweden about 1910 and acquired several mining claims. On one claim, at Squaw Creek, about eight miles from Chandalar Lake, he built a log cabin situated well above timberline. He laboriously dragged all the logs several miles to the site.

About 1949 he went to Fairbanks to work and make some money. He returned in winter, landing at Chandalar Lake without snowshoes. He had a terrible struggle in the deep snow in walking the eight miles home. When he got there he found a bear had torn a corner off his tiny cabin, which was now filled with tightly packed snow. It took days to shovel it out.

Ellis had a suspicious nature, harboring the notion that anyone who made friends with him was trying to steal his claims. I often flew over his cabin to check on him. He was so suspicious of those flights that he refused to leave the cabin to wave at me or to see who was buzzing him.

One winter I didn't see any tracks in the snow at the times I flew over his cabin, although a trace of smoke was usually curling from the stovepipe. This puzzled me, and I worried about the old man. One day that spring I landed on the shore of Chandalar Lake and found Ellis waiting for me.

He had broken his leg during winter and had somehow survived until spring. Then he had struggled eight miles on crutches he had carved to where I would find him and could fly him to the Fairbanks hospital. No one had been aware of his broken leg.

Doctors broke his leg a second time and reset it so it healed properly, permitting him to walk again. When his leg healed he returned to the wilderness to continue mining.

Red Adney, a longtime resident of Chandalar in the Brooks Range, shows off a lake trout he caught in Chandalar Lake. I'm standing in the background with my Cessna 180. JIM REARDEN PHOTO

I often wondered how Ellis survived such an ordeal. His everyday tasks such as preparing food were difficult even under normal conditions. How had he managed with a broken leg? How did he keep from freezing?

He never bothered to become a U.S. citizen and could not get the Territorial retirement pension. He eventually sold out for ten thousand dollars and lived frugally in Fairbanks for the rest of his days.

I marvel at the tenacity of such independent old-timers. Most sourdoughs like Amero and Ellis Anderson stubbornly refused to ask anyone for help or to borrow anything.

SOON AFTER ALASKA STATEHOOD in 1959, schools were established in most Native villages. This called for many new teachers, many of whom came from Outside (that is, from the South 48 states), attracted by generous salaries. Living quarters were generally provided in or near the schools.

A major drawback for many of these newcomers was the loneliness. They were strangers in a strange, unforgiving land. Often it was difficult for a city-bred teacher to adjust to life in a wilderness village. This was culture shock to which many simply couldn't adjust.

Icy winter temperatures usually discouraged visiting outside the village during the school year. Consequently at the end of the year, many of these teachers were anxious to get out of the villages and were ready to talk to anyone who would listen.

After one school year, the middle-aged teacher at the 63-person village of Hughes was anxious to return to her New York home. I made a special flight to Hughes because she wanted to get an early start.

That poor teacher talked incessantly during the flight. I nodded at what I thought were appropriate intervals, not paying much attention. At one point, I radioed a Wien DC-3 that was in flight, requesting that the plane stop at Bettles for the teacher. I had to plug in my earphones for the call.

When I clamped the earphones on, I apparently offended the jabbering teacher. She reached up and yanked them off my head, tossing them onto the floor, angrily announcing, "You're going to listen to what I have to say whether you want to or not."

Loneliness does strange things to some people.

ONE OF MY STRANGEST CHARTERS involved a former Alaskan, an engineer, who had moved to California and married a woman who was a judge. He was addicted to smoking, but so fearful of cancer that he had his mouth scraped periodically in an attempt to prevent cancerous growths. He decided to break himself of the tobacco habit by isolating himself in the wilderness of the Koyukuk.

He wrote, asking if I could fly him and his wife to a place where they would have no contact with people.

"No problem," I responded.

The couple arrived. The food and equipment they brought was more than adequate, so I agreed to fly them to a lake about sixty-five miles from Bettles.

Their gear was piled, ready for loading into the plane, when the man requested, "Andy, please go through our stuff to make sure there are no cigarettes."

His addiction was so serious he didn't trust himself not to hide cigarettes in his personal gear.

I flew the couple to the lake aboard a floatplane. Three weeks later when I landed, taxied ashore, and shut the engine off, the man leaped onto a float, opened the door on the passenger side, and climbed in. He was in a panic.

"Andy, you can shoot me, but I am not getting out of this airplane. I must have a cigarette. I'll pay for the charter to Bettles and return. Just don't ask me to get out."

The flight to Bettles Field took about forty-five minutes. I landed on the Koyukuk River as usual and taxied to the dock and shut down. The prop was still revolving when the engineer leaped ashore and ran full speed the half mile to the store.

His wife understood. No way could her husband stop smoking. When I flew the engineer back to the lake, the couple broke camp and returned to Bettles with me. He happily chain-smoked cigarette after cigarette as they waited a couple of days for the DC-3 to pick them up for the first leg of their flight back to California.

ONE DAY IN THE LATE 1950s, two middle-aged men arrived at Bettles on the mainliner flight from Fairbanks.

"We're prospectors," one of them said, "and we need to check out a few areas. Can you fly us?"

"Sure," I said. "How do you plan to pay?"

"I left a signed blank check with your traffic department in Fairbanks to cover all expenses," one of them said.

I called Fairbanks and was told they were holding the signed blank check and that I could fly the pair wherever they wanted to go.

I flew those two all over the country. I'd drop them off for a time, then pick them up and fly them to another spot they picked from a map. They had built up a couple of thousand dollars worth of flying and roadhouse bills when I received a frantic call from

Fairbanks. "Their check is no good; they don't even have an account at the bank. Provide no more services."

I later learned that these men had no prospecting experience. They were gamblers, scam artists. They had heard there was gold in the Koyukuk country and thought they would bluff their way into getting control of a gold mine and pay their expenses from whatever gold they might find.

A PECULIAR-APPEARING STRANGER climbed off the main-liner DC-3 one fall day and walked into the roadhouse. I suppose he would have been called a hippie in those days. A long scraggly beard half-hid his face. His greasy hair was tied into a ponytail and fastened with a piece of string. His clothing looked slept in.

He introduced himself to me. "I'm the teacher you've been corresponding with."

I could scarcely believe it. This was the schoolteacher I had been trading letters with for the past year? The one who had told me about his desire for a new lifestyle? His beautifully written letters bore little resemblance to this apparition.

"I want you to fly me to a place where I'm completely isolated," he said. "I don't want anyone near me. Nobody!"

I left the man—I'll call him Edward—at an abandoned mining airstrip where there was a warm cabin and plenty of nearby timber for firewood. The food he brought consisted mostly of pemmican—ground meat mixed with fat and other substances—that he had prepared.

Edward's last words as I left him standing on the little airstrip were, "I really want to be left alone."

I flew over the spot occasionally to check on him. After a couple of weeks, I saw he had put out a flag, the signal for me to land. He appeared to be safe and healthy, so I didn't land until I had the time a day or so later.

"Where in hell have you been?" he demanded to know. "I've been signaling you to land for days."

"You convinced me you wanted to be left alone," I explained, feeling a little guilty at having pushed him so far. I was certain he was ready to return to civilization.

To my surprise he handed me a list. "I'm sick of pemmican. Here's a grocery list. Bring the stuff as soon as you can."

It turned out Edward was sincere about wanting to live in the wilderness, and he stuck it out.

Within a year he had moved to a nearby lake, where he built himself a nice log cabin. He bought a trapping license and was working toward becoming a registered big-game guide. At that point his lovely wife and young son joined him. I hadn't been aware he had a wife and son.

Both Edward and his wife were ambitious, hard workers. They were diligently settling in to become longtime bushrats—the Alaska term for people who prefer the independent, isolated life.

It all ended abruptly one day in a tragic accident when Edward stepped into the whirling propeller of an airplane that was preparing for takeoff. He died instantly.

The body was returned to his home state and buried in the family plot. Soon after, recognizing the futility of trying to live deep in the bush without Edward, his wife and son left Alaska.

PROSPECTIVE BUSHRATS in all flavors arrived at Bettles Field. One March a middle-aged couple from the South 48, recently married, with the bride's two pre-teen sons by a previous marriage, arrived at Bettles on the Wien DC-3.

"We've read all about Alaska and have decided to live off the land," the man announced.

They had sold everything they owned, and with no outdoor experience had chosen Bettles in March as their jumping-off point.

They planned to live in a tent heated with propane and had brought a small tank of propane and a burner more suitable for cooking than heating. Their supplies included potato chips, soft drinks, candy bars, and loaves of bakery bread—food mostly unsuitable for a long stay in the wilderness. Their sleeping bags were summer-weight. They did not have an ax or tools of any kind.

Deciding to give them a taste of the land they had reached, and yet ensure their safety, I offered to loan them a cabin I owned at a lake within an hour's flying time from Bettles. Before taking off for the first of two charter flights to the cabin, I asked, "Do you have snowshoes?"

"No, we won't need them. We have galoshes," the man said.

Galoshes are worse than useless in the Arctic's deep snows.

I landed my ski plane on the lake, taxied near shore, and shut the engine down. The father opened the passenger side door and jumped into snow up to his hips.

"Think you could use snowshoes now?" I asked.

He looked thoughtful, but didn't reply.

That night the temperature dropped to 50 below. Fortunately for them there was a good supply of firewood, so they remained warm.

That family lived on the lake for some time, eventually moving into another cabin. I continued to fly supplies to them when I could conveniently veer from a mail flight.

About a year after these folks arrived, I was returning to Bettles with an empty plane. I decided it would be a welcome change for the family if I treated them to a weekend visit at the roadhouse, a hot bath, good meals, and a Saturday night movie. I landed and made the offer.

"There'll be no charge," I said.

"We'd love it!" the woman exclaimed.

"Me too," both boys chipped in.

"You three go ahead. I'll stay here and keep an eye on things," the man said, rather sourly.

I flew the woman and two boys back to the lake the following Sunday. Luckily for me I took a friend along for the ride. As I stepped out of the plane the man met me and poked a .38 pistol in my face.

"You're trying to steal my wife!" he screamed.

I'm being a gentleman when I say merely that the woman was past her prime. The accusation was absurd.

Luckily my friend was a diplomat who understood the seriousness of the situation. I was in a mood to do battle. I thought the pistol was empty and the nut behind it was waving it for effect. After soothing words by my friend and backup from the woman and boys, we all retreated to the cabin for coffee.

During our friendly coffee klatch I had a chance to examine the .38. When I broke the cylinder and saw six live cartridges, I turned pale. Anderson's rule No. 1 after that: "Never invite a lady to your home for a visit unless she's accompanied by husband."

OVER THE YEARS I became wary when I encountered folks who had been isolated too long. It seemed that places with the fewest residents seem to generate the greatest turmoil.

For example, hard feelings were the rule at Wiseman, on the middle fork of the Koyukuk, with a population of about a dozen old-timers. Few of these residents spoke to one another.

I quickly learned never to take sides, and I ignored all petty squabbles.

I landed on one regular mail run to Wiseman. While sitting in the plane waiting for the postmaster, a man walked toward the plane carrying a coiled blacksnake whip.

I didn't think he meant to use it on me, but I wasn't sure. Before I could latch the door on the passenger side, the man yanked it open and growled, "Where is that weasel-eyed sonofabitch? He told a goddamn lie about me, and I'm going to whip the bastard all the way to his cabin."

Fortunately the "weasel-eyed sonofabitch" had missed the flight.

JOE ULEN ENLISTED in the Army Corps of Engineers during World War I and was sent to Wiseman by the Army. He claimed the military forgot all about him, so he just stayed.

His knowledge of radio and communications led him to establish a private radio station that eventually was tied into the Territory-wide Alaska Communications System. Each day he sent twelve hourly weather reports to the CAA at Bettles, using dot-dash signals. Joe also mined for gold, at one time was Wiseman postmaster, and at different times served as U.S. commissioner.

Joe liked his liquor, but there was a problem in acquiring it. In Territorial days it was illegal to send spirits to the villages. When Joe and others craved refreshment, they ordered by mail or sent a telegram to Fairbanks. Joe, of course, had his own radio system and could order whenever he wanted.

The booze would be concealed in some sort of crate or container for shipment. Wien, of course, had no idea what was being shipped, and we couldn't search every package for illicit liquor.

I began to recognize the people who were regularly ordering booze. The regulars were always lined up waiting for their shipments.

I soon learned how to tell when Joe expected a shipment: He would get the word "calm" confused and would transmit it in his weather reports as "clam." On such days he'd invariably telegraph to the Bettles CAA station that it was "clam and clear"—regardless of how bad the weather might be.

I still shake my head over the many times I had to abort flights to Wiseman due to lousy weather because I believed old Joe's fake weather reports.

CHAPTER EIGHT

RUNNING OUT OF ALTITUDE, COMMON SENSE—AND LUCK

In 1950 Wien Airlines made a big mistake: It bought a Republic RC-3 Seabee for my use at Bettles Field.

This single-engine amphibian, produced shortly after World War II, was powered by a 215-horsepower Franklin engine mounted above the large cabin. For a brief time after the war, the Seabee was one of the few available new airplane models.

Its purchase by Wien was a mistake for many reasons. With its short wings and 1,950-pound empty weight, it was grossly under-powered, and it took forever to get off the ground or water. It climbed slowly. And it was unsuitable for cold weather.

That June I found myself with a pile of freight for the Eskimos who often camped at Anaktuvuk Pass in the Brooks Range. With summer daylight lasting around the clock, I chose to fly some of this freight in the Seabee at night when there was little wind and no turbulence in the mountains.

I was in the pilot's seat of a Cessna 180 at Bettles Field in this photo from the late 1950s.

Because of the Seabee's poor performance, I decided to lighten up on this flight by limiting myself to a half tank of gas and only five hundred pounds inside the roomy cabin.

Bob Matson, the CAA engineer who had loaned me lumber for the freight shed, asked if he could accompany me. He wanted to meet the Eskimos. At the time these people lived in caribou-skin tents, depended mostly on caribou for sustenance, and trapped and shot wolves for their furs and the fifty-dollar Territorial bounty. They existed much as nomadic Eskimos lived for centuries, except they used rifles instead of bow and arrow.

Against my better judgment I agreed to take Matson, despite the extra weight he represented. I expected that getting him to Anaktuvuk Pass would be the problem. I had it backward. As I learned to my sorrow, getting off the water at Anaktuvuk Pass was the problem.

Our midnight takeoff on wheels from Bettles Field was straight-forward, as usual requiring much of the runway. I flew up the John River and into broad Anaktuvuk Pass and soon located the Eskimo camp near a lake. I landed on the lake and Bob and I unloaded the freight. My decision to fly at night to avoid the wind now worked against me, for it was dead calm and the lake was as smooth as glass. A little breeze would have helped my takeoff.

Commonly on a small lake with a plane on floats, it's helpful to make a "step turn"— taxiing the plane at high speed on the pontoon step while turning, and building speed while circling. Speed produces lift.

But the Seabee floats were about two-thirds of the way out on the wings. Making a sharp step turn with the Seabee was impossible. I had no choice but to use the full length of the rather small lake with a straight-ahead takeoff.

I knew takeoff was going to be iffy, but there wasn't much I could do about it. I was unaware of the rocks the Eskimos had placed at the lake outlet. When they shot caribou swimming in the lake, the bodies drifted to the outlet and hung up on the rocks where they could be easily recovered.

The air isn't exactly rarefied in Anaktuvuk Pass, which is at 2,200 feet, but with the Seabee I soon became aware that every foot of altitude decreased its performance. With Matson aboard, I taxied to the far shore, pointed the plane's nose toward the outlet, and gave her full throttle.

It soon became obvious we were going to need all of that lake for takeoff, for the plane was sluggish. We were close to flying when the hull bounced across the rocks near the outlet.

That slowed us. Then in the narrow outlet stream, the wings crashed into willows that grew from the banks, and we came to an abrupt stop. The left wing float had been torn off, but the smooth hull of the amphibian skidded onto the bank without any other visible damage.

With the Eskimos pushing and full power from the engine, I ran the plane up on the tundra, where I lowered the landing gear. From there we worked it back into the lake for another try.

On a water takeoff the Seabee wallowed from side to side, with its two wing floats alternately in the water. They were there to keep the wingtips out of the water. I know, because I tried to take off without the missing float and nearly sank the plane. The wingtip without the float hit the water, brought us to an abrupt halt, and spun us in a tight circle.

If I hadn't shut the engine down promptly, I probably would have sunk that aluminum misfit right there. Bob Matson had to climb on the high wing as a counterbalance so I could taxi to shore.

I beached the plane and stood scratching my head until Eskimo ingenuity came to the rescue. One of the Natives sacrificed a dogsled runner, shaped it into a splint, and lashed it tightly into place. We reattached the missing float to the one-time sled runner with a tight lashing, and we were back in business. Or so we thought.

Closer inspection revealed a hole in the hull near the keel. I had nothing for a patch. The Eskimos were gathered around, lifting and pushing, listening and advising, and obviously enjoying themselves. Most of the talk was in their language, so I didn't know what was being said, but clearly such unusual, dramatic entertainment rarely came to these wilderness wanderers.

"If you had a boat, what would you use to fill the seams?" I asked one English-speaking Eskimo.

"Caribou fat," he answered.

One of the men brought me about five pounds of the stuff. We built a fire from dry willows and melted the fat, in which I soaked rags from the plane. While the tallow-soaked rags were still hot, I punched them into the hole.

It worked. This time, with two wing floats keeping the wings level, I got that doggy Seabee into the air and flew back to Bettles. As far as I know, the tallow patch was still holding several hundred hours of flight time later when the company, to my satisfaction, sold the airplane.

Noel Wien had two quotes he used to sum up pilot mishaps with an airplane. For minor events he'd announce, "It was just an old student trick."

But when an airplane was badly damaged, he'd say, "So and so is an awfully good pilot. He just managed to run out of altitude and common sense at the same time."

I had run out of common sense when I took a passenger with me to Anaktuvuk Pass aboard that Seabee. And that's what caused me to run out of altitude.

THE NEXT AIRCRAFT that Wien Airlines bestowed on me was little better. It was a used, three-passenger Piper PA-12 Super Cruiser with a 100-horsepower Lycoming engine.

This high-wing, fabric-covered model had few of the attributes of a good bush plane. Its biggest drawback was absence of flaps, so important for getting off the ground quickly and landing on short runways. Further, the cabin was small, with little room for bulky packages.

Though several engines of various horsepower were approved for that model, my plane came with the lowest-horsepower engine. It carried a useful load of about seven hundred pounds (passengers, freight, and fuel).

Nevertheless, with this misfit I did my best to please customers and to be an asset to Wien by accepting every job I was offered. On one operation for the U.S. Coast and Geodetic Survey, I flew for nearly eighty-nine hours over nine days and moved sixty thousand pounds of freight, grossing five thousand dollars. In 1950 that was more than the cost of the airplane.

One of my customers, a cantankerous miner, was preparing for the mining season. He had loaned a jackhammer to friends at another mine, and he asked me to fly it back to his mine. The mine using the jackhammer didn't have a landing strip, and I planned to land on a nearby river sandbar. However, the spring breakup of ice and snow was under way, and melting snow had raised river levels. All suitable sandbars were under water.

"Andy, where's my jackhammer?" came the radio query about once a day for a week or so. Each day I checked the sandbar I wanted to use, and each day the water was still too high. In the meantime, a member of an Air Force crew working in the area told me he would like to experience landing on a sandbar in a small aircraft. I welcomed the idea of taking this fellow along to provide him with the experience—and to get help in loading the heavy jackhammer.

Finally the river level dropped, leaving an island bar. I decided it was enough for a try, and with my eager Air Force helper I easily landed on the still-damp sand and gravel. I shut the engine down, climbed out, and walked up and down the bar. The longer I looked, the shorter it appeared. I began to have doubts about getting my no-flap airplane off safely carrying a heavy jackhammer and the husky passenger.

Under similar circumstances I sometimes placed a small-diameter pole across the far end of a runway, depending on it to bounce my plane into the air. If a plane is moving fast enough, once in the air it will usually remain airborne. But in this instance we were in the middle of a river with no usable poles in sight.

I decided to postpone hauling the jackhammer. Working the plane to the farthest end of the sandbar island until the tail wheel was in the water, I used full throttle to start our takeoff. The plane gathered speed slowly and the tail came up. The sand was soft and the tires sank deeply.

With the engine roaring and the wheels still firmly on the ground, we plowed into the river at the far end of the bar. The plane was knee-deep in the river when the prop struck water, abruptly stopping the engine. Slowly, the airplane flipped onto its back.

We swiftly struggled out of the half-submerged fuselage. We were soaking wet, but unhurt. I tied the airplane in place so the river couldn't carry it away and dug out the emergency gear. I was chilly and tired, but when my Air Force buddy suggested I join him in his single mummy sleeping bag, I politely declined.

When the plane flipped, a box of about two dozen blasting caps, intended for the owner of the jackhammer, was under the seat. None exploded when the plane cartwheeled, but we had to be careful where we stepped; blasting caps were scattered throughout the area.

Late that night the pilot of a passing helicopter spotted the upside-down plane, landed, and flew us back to Bettles. Eventually the river receded, I flew a couple of mechanics to the plane, they repaired it, and it was flown out and put back into service.

This was another case of "running out of common sense and altitude at the same time."

ALL PLANES FLYING IN ALASKA are required to carry emergency gear. I had been flying from Bettles Field for a couple of winters when that gear probably saved my life.

In winter I always carried one sleeping bag for every three passengers, an ax, snowshoes, and food enough to feed four people for two weeks. There were always tie-down ropes for the plane. Standard equipment for winter flying also included a plumber's pot (stove) with a gallon of white gas for preheating the engine, a canvas engine cover, and wing covers.

A plane parked outside overnight without wing covers can accumulate frost on the wings. Without wing covers I could spend up to half a day removing that frost before I dared take off. Frost on wings has killed many pilots. It doesn't take much to spoil airflow and reduce lift, and unwary pilots, thinking a touch of frost won't make much difference, can find themselves at the end of a runway in a stalled and wrecked plane.

The planes I flew at Bettles Field were maintained at the Wien main hangar at Fairbanks. After a regular inspection was completed

It took some unusual repair work to fix the damage after I tore a wing float off this Republic Seabee amphibian in a takeoff attempt near Anaktuvuk Pass in 1950. Eskimo ingenuity saved the day. They fastened the float back on the plane using a dogsled runner and rawhide and came up with caribou fat to patch a hole, and I flew the repaired plane home.

ANDY ANDERSON PHOTO

on my four-place Piper Family Cruiser one cold winter's day, I was flying a lightly dressed passenger the 180 miles back to Bettles. About halfway, as we flew over the Yukon River, I noticed specks of engine oil on the windshield. This wasn't particularly alarming, but it alerted me, and I watched the windshield and instruments carefully. After another fifteen minutes, the plane's oil pressure gradually started to drop.

The plane was on skis and I could have landed almost anywhere on the snowy land, and if I had been alone I would have set down to see what the problem was. But my passenger was not dressed for cold, and, fearing a landing might mean remaining on the ground, I continued flying toward Bettles. Oil pressure continued to drop. On final approach to Bettles I reduced throttle—and the engine quit. I was close enough to make a dead stick landing on the runway.

A pilot long remembers the sudden silence when an airplane engine quits abruptly in flight.

Examining the plane, Canuck and I found almost all of the engine oil had been pushed out through the seal around the pro-peller, congealing on the front and underside of the engine cowling and on the belly. We grounded the plane and asked Wien to send a mechanic from Fairbanks on the next mainliner DC-3 flight.

Thin-as-a-rail mechanic Aubrey Taylor arrived. Aubrey always wore the oversize, insulated arctic footwear known as bunny boots, and his friends called him Li'l Abner because of his resemblance to the comic strip character.

He puzzled over the problem, and it seemed he had it solved. I flew the plane on mail runs over the next few days. On some of the flights the plane lost a bit of oil. On other flights all seemed normal.

The weekend was nearing and Li'l Abner and I decided it was prudent to fly the plane back to Fairbanks where it could be worked on in the warm hangar. We tossed in an extra tarpaulin and checked the survival equipment before leaving Bettles.

I had mail to deliver to several villages on the way. At each stop we examined the plane's engine without finding any problem. So we were startled when the engine quit while flying above the Koyukuk River about fifteen miles north of Hughes, which was to have been my last mail stop. While gliding toward a landing I tried to restart the engine. The prop turned over slowly, but the engine didn't fire. I made a simple dead stick ski landing on the frozen river.

The whisper of skis on dry snow and the creaking of the plane—noises that cannot be heard over the hum of a running engine—were the only sounds. The plane stopped. Li'l Abner and I looked at each other.

"Now what?" he asked.

"We're about fifteen miles from Hughes. Feel like snowshoeing the rest of the way?" I asked, half joking.

"Nope," he said. "I'd rather have someone come get us with a plane."

"OK," I said. "Let's see if this thing will start. If it won't, we'll have to dig out the emergency gear."

The temperature on the ground was 35 below. Aloft a couple thousand feet, it had been much warmer—near zero. An aircraft battery will turn an engine over perhaps a couple of dozen times at 50 degrees above zero. In deep cold the number is cut at least in half, and the battery and engine must be warm. I hit the starter, and the prop turned over once and stopped. Our battery was done. There seemed no point in trying to hand-prop for a start since the engine had flat quit. We were stuck.

There are only four or five hours of decent daylight in midwinter in the Koyukuk, and a good half of that time was gone. Unless we were rescued soon, we knew we were in for a miserable night, so we put on snowshoes and went to work.

First we tied the plane down. Under each wing, using snowshoes, we dug through three feet of dry, powdery snow. With an ax I cut bridges in the river ice—two adjacent holes, with a tunnel between for a rope to pass through—for tie-down ropes. When the plane was tied down and safe from wind, we packed our survival gear into the shelter of the nearby spruce forest.

With the ax we cut firewood. We drenched the wood with gasoline drained from the plane, and it flared into a fine, leaping fire. It warmed our fronts, but our backs froze. After a time we would warm our backs and freeze our fronts.

We ate concentrated food bars, all the time listening for a rescue plane. Near dark I snowshoed back to the plane, peered at the windshield-mounted thermometer, and was startled to see the needle pointing to 50 degrees below zero.

It was too cold to expect a rescue plane.

We doubled our big canvas and slid our mummy-type sleeping bags inside, removed our mukluks, and crawled in for the night. Neither of us was really warm or comfortable. Sometime in the middle of the night Li'l Abner had a call of nature. He became so cold he couldn't move and I had to get up to go to his assistance. The poor skinny guy had no fat whatsoever to insulate him, and he was particularly susceptible to the cold. Nowadays we'd recognize such signs of hypothermia, but in the 1950s I don't remember hearing the term. At the time we were just plain, damned numbing cold.

That was one of the longest nights of my life.

The temperature remained at 50 below all the next day. We kept a fire burning, watching, waiting, and praying for someone with a plane to pick us up. We cut a lot of firewood. Dragging it through the snow warmed us a bit, and when the fire flared again we got warm. We rigged the tarpaulin as a reflector and stood between the tarp and the fire. That kept us reasonably comfortable.

We didn't even consider trying to restart the plane. I've seen big pilots hang their full weight from the propeller of a small plane at 50 degrees below without the propeller budging an inch. Until it could be repaired and thoroughly heated, that engine wasn't going to start.

As we kept our fire going, alternately basking and freezing, Li'l Abner and I dreaded another icy night in our sleeping bags. We considered several alternatives. One was to take turns keeping a fire going, which didn't appeal to either of us. Li'l Abner's bag was a double mummy, mine was a single bag. In the end we slid my single bag inside his double. With each of us in our own bag, we at least remained semi-warm, not in danger of freezing.

Wien pilot Bill English landed a company Cessna 170 beside our crippled plane late the second day to pick us up, despite the temperature of about 55 degrees below. I was never so glad to see an airplane. He flew us back to Bettles.

When the temperature moderated, mechanics were flown to the downed airplane. With fresh oil and preheating, the engine started and the plane was flown back to Fairbanks.

The airplane had been winterized earlier, but a newly hired mechanic, unfamiliar with cold-weather operations, had removed the insulation on the oil lines and oil reservoir. Water vapor escaping from the oil vent froze, plugging the vent. Internal pressure then pushed oil out the nearest weak point—the propeller shaft seal.

ADVENTURE ALSO LAY IN WAIT after I flew a prospector from his base at Wild Lake to the nearby John River, where he prospected for about a week. For several days before the day scheduled to pick him up, rain poured unceasingly, and the usually tranquil John River ran high and fast, with drift of all kinds swirling on the surface.

I flew over the prospector's camp, saw he was there, and landed against the current without difficulty, although I had to dodge chunks of driftwood. I taxied to the bank, cut the engine, stepped ashore, and tied my Cessna 170. We loaded his camp gear and prospecting tools, and I put him in the copilot's seat.

I cast off and we drifted into swift water. I elected to take off with the current to take advantage of the extra speed I would gain. The Cessna was planing on the step, but the river was quite crooked, and I lifted the right float out of the water to negotiate a bend. Around the bend we went, and I was ready to lift off when, with a bang, the left float struck a submerged object. This sent us tumbling.

Neither of us was hurt and we scrambled out of the plane, which ended up near the riverbank. I tied the plane to a tree to hold it in place until the river subsided.

The prospector said "To hell with flying. I'll walk home. It's a lot safer." He took off on foot, leaving me to wait for help.

When I didn't arrive back at Bettles as scheduled, Ed Klopp, the CAA station manager, took off to look for me in the T-Craft I had sold him. He spotted the wrecked Cessna almost immediately, landed, and flew me back to Bettles.

Another Wien company floatplane was flown from Fairbanks for my use. Mechanics ran a riverboat from Bettles Field up the John River to the wrecked plane, took it apart, and floated it back to the field. From there it was taken to Fairbanks to be rebuilt.

COPING WITH THE ARCTIC

In January 1955, with temperatures hovering at nearly 60 below, a radio call came from the village of Huslia, about 160 air miles downriver from Bettles. Could I fly a patient needing medical help?

Sidney Huntington, a trapper and gold mine worker, had been splitting wood. A sliver from a frozen spruce knot had flown up and embedded itself in his right eye. Fortunately for Sidney, the itinerant nurse Eunice Berglund happened to be in the village. She administered an anesthetic but couldn't do anything about the splinter.

Wien policy grounded all planes at 50 below. The main reason was concern for passengers. Repeatedly, lightly dressed passengers arrived at Bettles in winter on the mainliner DC-3 expecting to fly with me to a village, a mine, or a trapping cabin. Some were dressed as if they going for a short walk in a mild climate—about like I was when I arrived at Bettles Field for the CAA in 1947.

However, there was danger even for the Alaskans who were aware of the hazards of extreme cold. A light airplane engine cannot generate enough heat to keep the interior of a plane warm in the deep cold. When a plane is forced down in bitter cold, even with good emergency gear, the situation is life-threatening to everyone on board—including the properly dressed pilot.

Deep cold is insidious. It's a constant pressure, an inescapable menace. Toss a pan of boiling water into the air at 50 below and the water turns to vapor with a sudden whoosh, forming a miniature cloud. No water reaches the ground. Synthetic rubber tires, used during World War II, became solid at this temperature. Synthetic tire tubes crack and are virtually worthless in deep cold. Metal may shatter when struck. Spruce trees sometimes freeze and split with a report like that of a heavy rifle. So do logs in a building; I have often been awakened in the middle of the night by the rifle-like crack of a log in my Bettles roadhouse as the outside temperature plunged somewhere below minus 50.

Ice fog, which starts to form at about 40 below, is a nemesis for pilots, particularly around a busy airport. Aircraft and automobile engines spew moisture, which condenses into what in reality is a cloud, but is commonly known as ice fog. Ice fog at Fairbanks airports may reduce visibility to a quarter mile or less. Under such conditions flying stops.

It was near dark when the request came for the medivac flight for Sidney Huntington. I promised I'd make the flight the next day if the temperature climbed. But it was even colder the next morning. Nevertheless, in hopes that it might warm up I heated the engine of an airplane and kept it ready to go all day as I watched the thermometer.

Late in the day came another call from Huslia. Sidney was in great pain and urgently needed help. Could I please come?

I promised I'd fly to Huslia next day regardless of the temperature. I was up early, and Frank Tobuk and I had the airplane engine heated and ready to go by daylight. It was about 10 that morning before it was light enough to fly, and I took off and followed down the Koyukuk River to Huslia.

It is possible to fly in Alaska's arctic when ground temperatures are 50 below and colder only because of a temperature inversion; normally, it is much warmer aloft. With a ground temperature of 50 below zero I have often enjoyed a relatively balmy temperature of 20 or 25 below at 3,000 feet.

I radioed ahead and Huslia villagers were expecting me. Several people, all wearing fur parkas, including nurse Eunice Berglund, were standing at the end of the hummocky runway when I touched down. I kept the engine running, and Sidney, with a patch over his right eye, climbed into the plane and turned to me with a heartfelt, "Thank you for coming, Andy." He looked tired.

I took off and flew to the Alaska Native Service hospital at Tanana, 150 air miles away.

Sidney was in much pain; he had endured two nights and a day with nothing but pontocaine, an anesthetic administered through drops into the eye. The splinter was still in his eye.

Although it was colder than 50 degrees below zero at Tanana, visibility at the runway was good. No planes had been flying to create ice fog. Smoke from cabin stovepipes hovered over the nearby village, but I had no difficulty landing. A Jeep from the hospital was at the runway to take Sidney to the hospital.

"Don't leave, Andy," the doctor who drove the Jeep requested. "I'll take a look at Sidney's eye, but he may have to go to Fairbanks."

While I waited I refueled the plane from five-gallon cans I bought from the local distributor, pouring it in through a chamois skin in the funnel to filter out ice and dirt. I restarted the engine and idled it to keep it warm. In less than an hour the hospital Jeep was back with the doctor and Sidney.

"Andy, you've got to take him on to Fairbanks. I've called Dr. Fate. He's expecting you. I can't do anything for that eye here," the doctor said.

*Peaks of the Brooks
Range and a small
glacier filled the view
from my plane.*

JIM REARDEN PHOTO

Sidney climbed back into the plane. He had a fresh patch over his eye. I had to admire the way he tried to be good company even though he was in misery.

I was concerned that ice fog might make landing at Fairbanks difficult. The Tanana flight service said weather at Fairbanks was marginal, but it sounded like I could make it.

I flew the 130 miles to Fairbanks. It was midafternoon, near dark, and I had worn out the daylight hours while flying nearly 500 miles. Although at our 3,000-foot altitude it was a relatively warm 25 to 30 below, Fairbanks was reporting a temperature of 52 below. As we neared the town I could see it was mostly obscured by dense ice fog. I dropped to 1,000 feet, used flaps, and crawled along at about a hundred miles an hour, following the railroad tracks toward the airport as visibility decreased.

I called the control tower. They were expecting me, having been informed by the CAA of my medivac flight. Visibility was down to a few hundred yards when I landed at Fairbanks International Airport. An ambulance was waiting to take Sidney to Dr. Fate.

Dr. Hugh Fate worked on Sidney's eye all night, removing as much of the wood as possible. Sidney retained partial vision in the eye, but it gave him much pain. Six months later the doctor had to remove the injured eye, and since then Sidney has worn a glass eye. He went on to spectacular success in a business and public service career that in 1989 earned him an honorary doctorate for public service from the University of Alaska Fairbanks.

AT BETTLES FIELD, we never kept planes in a hangar. Mainly, we couldn't afford a hangar. Perhaps as important, I soon learned that when a plane came out of a warm hangar into deep cold, we sometimes had problems with it for a week or so. The pitot tube, which tells us the airspeed, would be frozen about half the time and I flew many miles without an airspeed indicator. It would be one thing after another as ice jammed this and that part.

Moisture condenses and freezes on an airplane when it is pushed from a warm hangar into deep cold. Practically every moving surface can freeze in place, including rudder, ailerons, elevators, trim tabs, doors, and cowling. If the plane is left in the deep cold long enough, sublimation will remove the ice, but that isn't much help if you want to fly the plane immediately.

We never greased flaps, elevators, and so forth with ordinary lube, which freezes up. Fittings that demanded grease received a special cold-weather grease.

After a day of flying in deep cold I always drained oil from the engine and kept it warm overnight in a heated building or on a stove. If that isn't done, a minimum of four hours are required to preheat a plane's engine enough so 50-weight aircraft engine oil will flow.

Sometimes I burned more gasoline preheating an airplane engine than I burned during a flight. The gasoline heater I used for preheating in my later years at Bettles Field produced 250,000 BTU and burned five gallons of gas an hour.

For emergency landings, winter was perhaps the best season for me at Bettles. After the first decent snowfall, a ski-equipped plane can land almost anywhere on rivers, lakes, tundra flats, and ridgetops.

I learned to be cautious on lakes, however, where overflow could be a problem. This occurs when the lake is frozen and a heavy burden of snow falls on it, pushing the ice down and allowing water to flow atop the ice, but under the snow. When I wanted to land on a snow-covered frozen lake, I learned to drag the plane's skis on the snow without putting full weight down. Two or three drags across an area would compress the snow and reveal water as dark streaks in the ski tracks. Water and winter temperatures are a dangerous mix. If no water appeared, it was generally safe to land, provided the ice on the lake was thick enough for support.

Spring breakup of the ice was always a nightmare. In my early years at Bettles Field most villages I served did not have aircraft runways. On many occasions I needed three types of landing equipment to complete a mail run. I flew as long as I could on skis, and when snow was gone I landed on wheels at villages with an airport—once the runway was dry. As spring advanced I landed on rivers or lakes with floats. Almost all Koyukuk Valley villages are on the banks of the Koyukuk River.

It was costly to have to use three types of landing gear in order to complete our mail contract. Sig Wien realized this. He also felt that every sizable village we served should have an airport large enough to accommodate our largest plane year-round. Therefore, Wien built good landing fields at Venetie, Kobuk, Arctic Village, and Anaktuvuk Pass. The company installed two-way radios in these and other villages, establishing a radio network that made it possible for bush villages to communicate with Wien, with each other, and with Fairbanks twenty-four hours a day. Also, in cooperation with the Alaska Department of Aviation, Wien built fields at Huslia, Hughes, Allakaket, and Kiana.

These runways improved the standard of living in these villages. Residents could now come and go year-round for work or medical reasons. Mail service became consistent, and supplies could be delivered any time.

PLANNING A YEAR OR MORE in advance was required for most

fieldwork in Alaska's Arctic by oil companies, seismic operators, U.S. Geological Survey field parties, and others. On one occasion Wien Airlines contracted to stockpile aviation fuel and other supplies for a survey party that expected to arrive in the summer with a bevy of helicopters.

Peters Lake, four and a half miles long and a mile wide, on the north slope of the Brooks Range, was selected as the site for storage of fuel and supplies. In winter, of course, it is frozen over and covered with deep snow. To store the supplies, it was necessary to wait for the snow to melt as spring weather warmed. This produced a large volume of water, which drained off around the edges of the lake. The four- to five-foot-thick ice that supported the snow now broke loose from the shoreline and lifted. For some weeks, the bare floating ice made a perfect landing place for a large plane.

For hauling gasoline and other freight, Wien had several twin-engine Curtiss Commando C-46 airplanes. That spring the company ran a little behind on delivering freight to Peters Lake because of the enormous quantity of gasoline and other supplies involved. As spring came on, surface ice on the lake became a little mushy during warmer days, but it froze solid overnight.

On the last flight for the season, a C-46 with thirty-five drums of aviation fuel on board made a perfect landing. The engines were slowing to a stop when suddenly the pilots heard "click-click-click" as tips of the propellers hit ice. The airplane was gradually sinking. Within hours, the 29,000-pound airplane was supported only by its belly. Both wheels were submerged in about five feet of rotting ice.

The ice was too soft to allow the plane to be jacked up. A good crust formed nights, but daily, before noon, water started to flow atop the ice, and the plane kept settling deeper.

Fearing it would go to the bottom, the company had mechanic

Li'l Abner remove radios and instruments. In the meantime, with light planes the company flew in about fifty empty oil drums, which were fastened to the plane for flotation.

The ice finally melted and the airplane floated on the empty drums. With an outboard motor, it was pushed to the shallow end of the lake. The landing gear was already down, and with the help of several cable winches, the C-46 was pulled up on the beach.

Li'l Abner and another mechanic reinstalled the radios and instruments and drained water from the engines. The north beach of the lake was fairly smooth and suitable for takeoff. When all was ready, pilots warmed the engines, said a prayer, gave her full throttle, and yanked the big plane off the beach as soon as it had flying speed. They flew to Fairbanks, where the plane continued to work for Wien for many years.

VILLAGERS ALONG THE KOYUKUK were always skeptical of newcomers, especially white newcomers. Too many whites, including missionaries, had taken advantage of them. After I had flown from Bettles for a couple of years, I found that much of the initial skepticism about me had disappeared.

As I grew closer to the people, I found that serving their needs for the sake of money had become a secondary motive for me. It gave me a great sense of well-being and pride to be considered one of the locals who, by providing aerial service, was in a position to improve their lives.

To gain and keep the confidence of these wonderful people I felt it necessary to do my very best to deliver the mail and freight as scheduled, regardless of weather. For this reason I sometimes pushed weather more than I should have.

That was the case one midwinter day. I shouldn't have left Bettles in the first place, but I had managed to sneak into Hughes on a mail flight in advance of a forecasted snowstorm. As I left Hughes and headed home, snow reduced visibility to something under three hundred feet. With skis in landing position, I flew slowly within the confines of the river. Tall spruce trees thrust above me on both sides. An increasingly strong wind rocked the plane. Visibility decreased. Soon heavy snow built up in the engine air intake and the engine overheated and quit. Fortunately I was on a straight stretch of the river and made a simple dead-stick landing.

Once landed on the river ice, I tried to clear the air intake. While I worked, snow fell so heavily it was obvious I couldn't fly, for visibility was down to a few feet. A strong wind pushed me and the airplane across the river's glare ice. The wind also cleared the snow from the slick ice.

Since I didn't dare try to take off, the airplane had to be tied down or the wind would flip it or drive it against a bank. With an ax I cut a bridge in the ice to run a tie-down rope through. It was almost finished when the wind suddenly shoved the airplane about ten feet—far enough to make the bridge useless.

Again I frantically chopped another ice bridge, only to have the wind skid the airplane back another ten feet. This routine continued for nearly an hour as I chased the plane several hundred yards down the river.

The wind was now howling and the plane, its nose pointed into the wind, was blown steadily backward. At a bend the wind shoved it toward trees on a low riverbank. If I couldn't stop it, the plane was going to blow into the trees and likely be damaged.

Desperate now, I remembered a trick I had once seen an Eskimo use. I unloaded mail and freight, piling it on one of the skis to hold that side temporarily. I rushed around to the other side of the plane and with a few swings hurriedly chopped a simple depression in the ice. Laying the end of the tie-down rope in the depression, I urinated into the depression. It was 40 below and the urine froze immediately, holding the end of the rope in place. I hastily tied the other end to the ring on the top end of the lift strut. I duplicated this process on the other side. Both ropes held, and the plane stopped sliding.

Reloading the freight and mail, I made myself comfortable inside the plane, protected from the howling wind. My emergency gear included a small gasoline heater. It folded to the size of a medium-size camera, yet it produced a tremendous amount of heat and the inside of the plane was soon cozily warm.

Bouillon and hot tea made from melted snow bolstered with pilot bread made a skimpy meal. I melted more snow and poured the water atop the ends of my tie-down rope where it promptly froze, reinforcing the frozen urine. At dark, wearing my down-filled parka, I crawled inside a down-filled sleeping bag in the back of the plane for what I hoped would be a good night's sleep

The tie-down ropes were frequently tested by powerful wind gusts, and I slept fitfully, for in strong gusts the plane lifted off the ice and flew itself. It was prevented from flipping over only by the ropes.

The ropes held. Next morning with the falling snow and wind ended, I cleared the engine intake, preheated the airplane with my plumber's pot, and flew home to Bettles.

KOYUKUK MINERS

One summer day a man arrived at Bettles Field on the mainliner DC-3, stalked into the roadhouse, and stood waiting until the plane was gone. "I'm from Anchorage," he told me. "I want you to fly me to a gravel bar where I can make wages panning gold."

"Where do you want me to take you?" I asked.

"That's up to you. You probably know the best place. Just take me where I can make good wages panning."

"But I don't know of any place where you can do that."

He became angry. "You must. You live here. You fly gold miners all the time."

I had a tough time convincing him. He only half-believed me when I said, "If I knew of any such place I'd probably be out there with a drag line myself making the gravel fly."

For about a century, gold mining has been important in the upper Koyukuk River valley. From 1900 to 1930, 5 million dollars worth of gold was mined there, most from the headwaters about seventy-five miles upstream from Bettles. Perhaps ten times that amount came out during the 1950s, when modern tractors and other heavy equipment became available.

While I was at Bettles, one mine was a large floating dredge that in 1955 was transported piecemeal up the Koyukuk by riverboat. It scooped gold-laden gravel from sixty feet below the surface.

That dredge alone reportedly dug out more than $20 million in gold within about a decade.

Some mines were family-owned and family-run. Some miners were loners who never revealed how much gold they dug each year and who disappeared to the south in late fall. I often imagined some of these characters spending their winters lounging on some tropical island surrounded by bevies of beach-loving beauties.

The Yugoslavia-born brothers Sam and Obrien Stanich, bachelors, lived and mined on Porcupine Creek. Not only were they efficient, hard-working miners, but they also raised vegetables that were the envy of the Koyukuk. They specialized in crops that could be planted early and harvested early because of the short season. Of course, their potatoes, cabbage, lettuce, carrots and turnips were flooded with sunshine practically twenty-four hours a day. And my, how those vegetables grew!

I saw potatoes that weighed three and four pounds, cabbages of thirty-plus pounds. From cabbages they made wonderful sauerkraut. It was so good that I bought a wooden keg of the stuff from them every year. We also fed our roadhouse guests many other products from the Stanich garden.

In searching for gold, the Staniches dug their shafts by hand through permafrost in search of bedrock, which was usually thirty or forty feet down. The frozen ground had to be thawed every foot of the way by building wood fires atop it. They had dug so many shafts and built so many fires that there were virtually no trees for several miles around their claim.

They stored their vegetables in one of the shafts. It served as a root cellar, keeping the produce moist and crisp.

How much gold did I fly out of Koyukuk mines? I have no idea.

I can say that from Utopia Creek and Hog River, I flew gold out about every three weeks during summer. It always took three or four strong men to lift the strongbox into the plane.

It was commonly believed that at Myrtle Creek, $100,000 worth of gold was mined each season. When that figure was reached, the mine shut down for the year; the owners regarded their claim as the best bank ever.

Flying miners and their supplies and equipment was an important part of my business. No roads existed in the region other than frozen winter trails where crawler tractors and huge sleds brought in equipment that couldn't be airfreighted.

The major mines had their own aircraft runways, but at many small operations I had to land on gravel bars or with a floatplane on a nearby river or lake.

ESKIL ANDERSON, A MINING ENGINEER who lived in the state of Washington during the winters, had a hard-rock claim in the Chandalar country of the upper Koyukuk. At his mine, rock containing the ore had to be blasted out, then crushed. Gold was recovered from the crushed ore.

One summer I flew Eskil's son and another young man to the mine, where they blasted rock out and stockpiled it for later crushing. Both were freshmen in college. Like so many young people, they wanted a little excitement in their work. So they made a game of planting dynamite with short fuses. When the fuses were lit, they ran pell-mell for cover.

One day when the dynamite blew, they were still running. A rock about the size of an orange caught young Anderson in the small of the back. It knocked him down and he stayed down. Anderson's

The Wien Cessna 180
is being readied for
departure at the
Koyukuk River
floatplane landing
at Bettles, 1955.

friend called me on the radio, asking me to fly the injured man to a hospital.

The Anderson mine is at an elevation of 3,500 feet in a crevice between two mountains. The dirt runway of 1,500 feet is approachable from only one direction, regardless of wind direction. At times I couldn't land there when the wind was wrong.

Flying a Pilatus Porter, an eight-place turbo-powered airplane, I arrived within an hour of the call when, luckily, there was virtually no wind. I touched down, taxied to the camp, stopped the engine and trotted to the cabin where the injured youngster lay in agonizing pain. There were no pain suppressors at the mine or in my plane.

With support from me and his friend, he managed to walk the short distance to the airplane. I removed all the seats and made a pallet out of sleeping bags and wing covers for him to lie on. I placed him next to my seat so I could keep an eye on him. Then I flew full throttle to Fairbanks.

Serious injury can bring on shock, and I felt that warmth might offset it. Even though it was summer, I turned the cabin heat to maximum. I made a point of talking frequently to the young man, trying to keep him alert. He realized his predicament and made a great effort to talk with me and remain calm on the two-hour flight. I radioed ahead and an ambulance was waiting at the international airport.

A few weeks later I received a letter of thanks from Eskil. His son, fully recovered, was anxious to get back to the mine. I doubt if he ever again gambled with short dynamite fuses.

OLD-TIME PROSPECTORS were scattered throughout the Koyukuk Valley. Many lived alone, and they were a crusty, independent lot.

Although few of these miners were on my scheduled mail runs, on occasion I stopped to check on these aging gold hunters.

Sometimes I was able to help them in little ways, but usually I received more than I gave. In my early years at Bettles there were virtually no aerial maps of the country. The old-timers were a library of information about the country, its history, and ways to cope with hardship. I soaked up every helpful nugget of wisdom.

One sourdough I occasionally visited was Frank Bishop. In his late 60s, Frank lived alone at Wild Lake about forty-five minutes flying time from Bettles. Summers he mined and prospected, and winters he trapped for fine furs. His expenses were minuscule, and he had a reasonably stable income.

One summer day as I flew over, I saw him near his cabin and decided to land to say hello and to see if all was well. I touched down on the lake and taxied to his cabin.

"Glad to see you, Andy," he called, when I climbed out of the plane. "Come in and we'll have some coffee."

I didn't really want to stay that long, but he was insistent. In his neat cabin he fired up the woodstove and put water on as we discussed weather, some caribou that had recently drifted near, and late news. Like most loners who live in the Alaska bush, he had a shortwave radio and kept up with the news; he knew more about what was happening in the world than I did.

As he poured coffee he commented, "Andy, I've got a strep throat and I need a shot of penicillin. You'll have to give it to me."

"Who, me?" I burst out, horrified. "I'm no doctor. I've never given anyone a shot. I can't do it."

"Who else can I ask?" he said, eyebrows raised, looking me in the eye. "You're it. I haven't seen anyone else in weeks."

Reluctantly I agreed to give him a shot in the buttocks. He

dug out a hypodermic syringe, with needle attached.

"The needle will have to be sterilized, won't it?" I commented.

"Naw. It's OK," he responded.

I drew the line at that and insisted that I would give him the shot only if he sterilized the needle first.

"Well, then, OK," he finally agreed.

The coffee pot was perking, so he removed the needle from the syringe and dropped it into the boiling coffee. He then dug some penicillin out of an old beer case containing an assortment of dried furs. I could hardly believe what I was getting myself into.

He lifted the needle out of the boiling coffee with a pair of pliers and stuck it on the syringe, filled the syringe with penicillin, handed it to me, then unbuckled his belt and bent over.

I jabbed the needle in and pressed the plunger, emptying the milky-appearing stuff into his behind.

When I returned to Bettles Field I told a visiting army doctor what I had done, and asked what complications might occur.

"Some people are allergic to penicillin," he said. "You should have stayed for an hour or so to see how the stuff affected him. A bad reaction can be life-threatening."

I hardly slept that night. Early next morning I flew back to Bishop's cabin. He was surprised to see me again so soon, but welcomed me as usual.

"Andy, I wish you'd been here last night. For some reason I broke out in a sweat for a while. Then suddenly I had chills and damned near froze. Red blotches broke out all over my body. I didn't know what was happening to me. It's funny though; I'm OK today."

"You're allergic to penicillin," I told him emphatically.

"No, not me. I've had lots of penicillin shots. It was something I ate."

He was OK, so I left. A few days later when I was in Fairbanks, I visited every drugstore in town and warned the pharmacists never to send an order of penicillin to Frank Bishop at Wild Lake.

ANOTHER OLD-TIMER I GREATLY ADMIRED was Dennis O'Keefe, born in 1904 in County Limerick, Ireland. The big powerful Irishman had once been a professional boxer and in his youth was a sparring partner to Jack Dempsey. When I first knew him, O'Keefe thought nothing of snowshoeing through deep snow all day carrying a sixty-pound pack.

He built a cabin in a narrow pass everyone called Denny's Gulch.

There he sank a 40-foot hole to bedrock and found no gold. He then dug two holes 86 feet deep. For each eight inches of hole he had to build a wood fire, wait until the permafrost thawed, climb down, fill his bucket, raise it to the surface with a windlass, drag the bucket away from the hole, dump it, then lower it back down, and repeat the process. Each time he hoisted the bucket he got one cubic foot of dirt. He might have to climb the ladder down the hole twenty times to remove the dirt from just one eight-inch thaw.

It was difficult to fly through Denny's Gulch because it was narrow and funneled the wind, creating high wind speeds, often with violent turbulence. When it was safe, I flew over Denny's cabin and watched for smoke from the chimney or fresh snowshoe tracks—indications that he was all right.

In early 1957 I flew through Denny's Gulch about three times over a period of less than a month without seeing any sign of my sourdough friend. I became concerned and asked his nearest neighbors, thirty and forty miles distant from his cabin, if they had seen him. None had.

This was about the time of the uranium craze, when every prospector seemed to be searching for this mineral. The previous year Denny had found a deposit he suspected might contain uranium, but after many tests it proved to be pure sulfur. While making his tests he had inhaled sulfur fumes and became quite ill. He had struggled across country about fifteen miles to Big Lake, and I had flown him to Fairbanks for medical care. He recovered from that ordeal and returned to his mine.

Now really concerned, my investigation intensified. I made stops at half a dozen places at varying distances from his cabin. No one had seen Denny.

I asked about him at Big Lake, but he had not been seen. Then I flew for a few miles along the trail he would have taken to the lake. Suddenly I saw him lying, seemingly relaxed, beside the trail. His favorite dog, Bill, was fastened to his belt with a chain.

Denny was dead. Evidently he had been there for some time, for the poor dog had to devour much of his body to live. The Territorial police speculated he had again accidentally poisoned himself and was walking to where I could fly him to the Fairbanks hospital. This time he hadn't made it.

I flew his body to Bettles, built a beautiful coffin, and stored the coffin with him in it in a cold shed for the balance of the winter. I informed his only living relatives in Ireland. In the spring when the ground thawed sufficiently, we held a simple service and buried him in Alaska's wilderness, not far from our roadhouse at Bettles.

Sadly, because he had been forced to feed on his dead master to survive, Bill, a gentle, faithful animal, was destroyed. No one wanted to adopt him.

AN ELDERLY MINER named Arvid Repo was among the people I admired and liked. One late October he shut down his mine for the winter and flew to Fairbanks to relax for a few days and purchase his winter supplies. When he returned to Bettles on the Wien mainliner DC-3, I loaded him and his winter supplies into a plane and flew to the winter cabin at his claim.

"Boy, I'm glad to be getting home, Andy," Repo said cheerfully. "Let's just pile this stuff up here, and you can be on your way."

I was on my knees in the back of the plane, handing him boxes and bags, and he was chattering away happily when, suddenly, he became silent. I thought he had walked to his cabin. When he didn't appear for several minutes, I crawled out of the plane to look for him. I was shocked to find him lying motionless amidst his pile of winter supplies.

He had no pulse, and I failed in my attempts to revive him. I was badly shaken by this tragedy. I informed the State Police, and a friend of Repo's from Wiseman came to stay with the body until the police could come to investigate. The police eventually flew the body to Wiseman, where Repo's friends built a wood casket and buried him in the pioneer's cemetery.

FLYING THE MAIL

My once-a-week mail flights to the old mining camp of Wiseman often provided unexpected adventures. I never knew who was speaking to whom in that badly divided village of nine or ten persons. During my time on the Koyukuk, Wiseman was always a hotbed of post office problems. Regardless of who was postmaster, or how mail was handled, there were always complaints.

At one time the postmaster, a crusty old-timer preparing to retire, wanted to keep the position in the family and recommended his daughter as his replacement.

Another strong-willed, opinionated old-timer, a frequent complainer, applied for the job. There was no love lost between the two men. Each had a pet name for the other. The applicant called the postmaster The Wretch, while the postmaster called the applicant The One-Eyed Polecat.

Thinking it might end complaints, postal officials appointed the One-Eyed Polecat to the postmaster's job. This set off another, louder, round of complaints from Wiseman residents.

The Wretch wrote his objections to the postal inspector, the

In the cabin of a Wien Cessna 180, I'm getting ready for a flight. The year is 1955. JIM REARDEN PHOTO

Pilot Richard Wien delivers mail and freight at the inland Eskimo village of Anaktuvuk Pass in 1955. Airplanes were still a novelty, and most villagers gathered around when the Wien Airlines mail plane arrived. The burlap bags in center rear contain Eskimo-tanned wolf and caribou skins. JIM REARDEN PHOTO

governor of Alaska, Alaska's congressman, and the president of the United States. Shortly afterward, a government plane landed at Bettles with a handful of bureaucrats aboard, including representatives of the post office, the governor's office, and several officials of lesser stature. "Come on, Andy, this involves you too. You're going to Wiseman with us," the postal rep insisted.

"How am I supposed to be involved?" I asked. I wanted nothing to do with the quarrels at Wiseman.

The rep handed me a letter of complaint about the appointment of the new postmaster. I had difficulty keeping a straight face as I read it. It went something like: "When the One-Eyed Polecat that you have made the terrible mistake of naming as the new postmaster passes my home with his dog team on the way to and from the airport to pick up the mail, he points his posterior toward my house and wiggles it in a vulgar and obscene manner."

"Aw, that's just a typical letter from Wiseman," I commented, handing it back. Later I realized that after he had read it, the postal inspector's curiosity must have been aroused; my guess is he decided he had to meet the man who could write such a letter.

The bureaucrats visited Wiseman (without me) and met individually with the residents. A group meeting, of course, was out of the question; no one would have attended. The problem was solved by temporarily eliminating the Wiseman post office. This meant that I, the carrier, was supposed to personally deliver mail to each person in the village. I managed to get the two main factions to agree to accept their mail in bags, one bag for each faction.

Richard Wien, son of airline founder Noel, was then flying with me from Bettles to gain bush experience. One day I divided the mail for Wiseman into two sacks and gave them to Richard to deliver. I failed to explain the situation.

Richard flew to Wiseman and handed a rep from one of the two factions both sacks. Shortly the man handed one sack back. Richard thought it was outgoing mail. When he arrived back at Bettles I asked, "Richard, how come you brought half the mail back?"

He had to make the hour-and-a-half round trip to Wiseman again to deliver that sack of mail—to the "other" faction.

This silliness didn't last long. Another postmaster was soon appointed, making my life a little easier.

FLYING THE UNITED STATES MAIL from Bettles Field on a five-day-a-week schedule was an important part of my business. Passengers and freight were included on these flights.

Flight 801, made twice a week, included stops at Bettles Village, Allakaket, Hughes, Utopia Creek (an Air Force base), Hog River (a gold mine), and Huslia. All other runs were once a week, including Flight 803 to Wiseman, Flight 807 to Anaktuvuk Pass, and Flight 805 to Big Lake and Chandalar.

Wayne ("Red") Adney, a tough, one-time member of the famed Alaska Scouts of World War II (also known as "Castner's Cutthroats"), and his personable and outspoken wife Capitola ("Cappy") lived at Chandalar, the last stop. Red, who prospected for gold, trapped in winter, and guided nonresident big game hunters in spring and fall, was gone much of the time. Mail duties were left to Cappy.

To promote friendship and to catch up on local news, Cappy never failed to have a tasty lunch ready for me on mail days. This was fine with me—I had a good meal and at the same time I brought her up to date on local news, which I tried not to think of as gossip.

Sig Wien often visited Bettles Field to escape the stress of the main office and found relaxation by flying some of my mail routes. One day the run happened to be to Chandalar Lake.

I should have warned him. Sig wasn't familiar with the meal/gossip sessions Cappy engineered. He landed at Chandalar Lake and taxied the plane to the beach, climbed out, and unloaded a sack of mail. Cappy came out of the cabin to meet him.

"Well, Mr. Wien! Nice to see you. Come in. Lunch is ready."

Sig Wien was a taciturn, slow-talking, shy, deeply devout man. He knew Cappy's husband Red was not there, and he had no intention of eating lunch with this strange lady.

"Thank you, no. I have to get going," he replied, politely.

Cappy was not to be thwarted. "I insist. I've fixed a nice lunch. Andy always eats his lunch here on mail days," she said, still being polite.

"That's awfully nice of you," Sig said, "but I just can't. I don't have time today."

This give and take continued for several minutes. Finally Cappy had enough. She blurted, "Now Mr. Wien, lunch is ready. Andy always takes time to eat. Get your ass out of that airplane and up to the house and eat your damned lunch. If you don't I'm going to grind it up and feed it to you as an enema."

Mr. Wien lunched with Cappy.

On his return to Bettles Field I had a feeling that all hadn't gone as Sig expected. After I probed for a couple of minutes, trying to learn what had happened, he finally allowed, "My, the woman at that station is sure outspoken, isn't she?"

My curiosity got the better of me, and I flew to Chandalar Lake to learn the details. Cappy was still giggling. "I didn't mean to embarrass the poor man, but dammit I had a nice lunch ready, and I wanted him to eat it," she said, after confessing what she had said to Sig.

No harm was done; Cappy had simply offered the normal bush Alaska hospitality to a shy man. Sig didn't have a chance.

IN 1948 WHEN WIEN AIRLINES agreed to use Bettles as a distribution point for mail, freight, and passengers, the mail for the village of Bettles didn't amount to much, nor did it for the other post offices up and down the Koyukuk Valley.

However, traders in the region soon learned that it was cheaper to send and receive groceries by parcel post than by airfreight. It wasn't uncommon for me to receive tons of mail in the form of groceries—fifty cases of canned milk, fifty sacks of flour, and so forth.

Technically the mail was the responsibility of Jim Crouder, the Bettles Village postmaster. But with adequate safe storage for mail at Bettles Field, I was reluctant to fly it the five miles to Bettles Village, leave it, then pick it up later to fly to other villages.

While at Bettles on one of his regular scrutinies, the postal inspector for the region learned of the situation. He confronted me with Crouder's concerns.

"Andy, the mail you have piled up here is supposed to be under the jurisdiction of a postmaster. We have to solve this problem, now," he said.

"OK. What do you want to do?" I asked.

"Raise your right hand and repeat after me," he replied.

He then administered the oath of office establishing Bettles Field as a new post office, with me as postmaster. "Now you're responsible for all transient mail passing through Bettles."

He was one bureaucrat who knew how to instantly solve problems.

The aerial mail routes I established in the Koyukuk Valley were flown with no radio or navigation aids. Weather was often lousy.

Eskimo belles of Anaktuvuk Pass pose in front of the weekly mail plane in the mid-1950s.

ANDY ANDERSON PHOTO

Regardless, after a time the villagers not only expected, but demanded, that mail be delivered as scheduled and on time.

Sometimes weather was bad at Bettles Field at the same time it was good at one or more of my stops. Often villagers weren't aware of this and scolded me when I missed a mail run due to bad weather they didn't experience. I had to be careful to not allow this pressure to push me into unsafe flying.

Nevertheless, I flew hundreds of trips with virtually no ceiling and little visibility. The Koyukuk River is bordered by a spruce-birch forest, and it made a perfect route for the southern mail run because most of the villages are situated on the riverbank. For bad-weather winter flights I kept the airplane's wheel-skis in the ski position, lowered half flaps, and set up a cruise speed of 80 to 90 miles per hour, flying almost at treetop level and followed the winding river sometimes for the entire mail run. The clean, snow-covered, and frozen Koyukuk River was always visible between the dark boundaries of the spruce forest. After a time I had virtually memorized every river bend. This route consumed about four hours of flying time, after which I returned to Bettles by the same route.

I often encountered fog—clouds that reached the ground. Rather than turn back, when on floats I sometimes landed and taxied at 40 or 50 miles per hour along the winding river until I passed the low clouds, then took off to continue the flight.

IN THE LATE 1940s when I started flying to the village of Anaktuvuk Pass, the Eskimos there were nomadic, following the restless caribou herds of the region, depending upon the animals for virtually everything they needed. (Anaktuvuk in the Eskimo language means "place of many caribou droppings.") On early mail runs in winter I commonly circled to pick up their sled trails, then traced the winding tracks to where the people were camped.

They lived then in igloo-shaped caribou skin tents. Wood for heat was a luxury except for the few dried willows branches they could find in and around the treeless pass. Spruce firewood had to be hauled by dogsled for many miles, for the pass is well above timberline.

Turbulence was the main problem on my flights to Anaktuvuk Pass. Many times as I neared the mountains, my plane shuddered and shook so violently that I retreated home to make the flight another day. Landing on a lake at Anaktuvuk was seldom a problem—the wind was normally so strong I rarely traveled more than the length of the aircraft after touching down.

I always suspected the weather bureau had Anaktuvuk Pass in mind when it came up with the descriptive term "flowing snow," because the snow there always seemed to travel horizontally.

These Eskimos told me they needed six caribou a day for food for the 90 to 110 people in the band. No part of a caribou was wasted. Even the hooves were made into snow goggles. Their clothing, which the women sewed from skins they tanned, consisted of two layers of caribou skin, one of young short-haired skins with the fur turned toward the body, the other of longer-haired skins with the fur on the outside. Their wolf fur parka ruffs were spectacular and beautiful. The long guard hairs were arranged so they framed and protected the face from cold and wind. They surely needed this, for the wind in the pass is almost constant.

Anaktuvuk Pass, at an elevation of 2,200 feet, is an awe-inspiring place. Water from one lake in the pass flows south into the Yukon River, while water from the next lake, a mile or so north, descends into the Arctic Ocean. Magnificent 5,000- to 6,000-foot-high mountains rise on both sides of the pass.

Because of their long isolation, the residents were unusually susceptible to colds, influenza, and other diseases, and I had to be very careful not to take anyone with a bad cold or other symptoms there.

The main source of dollar income for these hardy people was wolves they trapped and shot. Each brought a fifty-dollar bounty payment from the Territory (and after 1959, the State) plus the thirty to fifty dollars the fur was worth. (Alaska stopped paying a wolf bounty in 1970.)

The postal department insisted the Anaktuvuk people establish a main camp so mail could be delivered to them regularly, and they chose to make the pass their permanent home. From caribou skin tents, their housing evolved to sod huts, then to plywood frame huts. Later, the children were sent to a boarding school. Despite their isolation and the difficult arctic conditions, I found them to be a happy, fun-loving people.

Life in Anaktuvuk Pass changed a great deal more in later years as it began seeing its share of the wealth from oil in the North Slope Borough, in which the village lies. As the twentieth century ended, the village had a fine year-around aircraft runway, a large modern school, permanent homes, and electricity. But caribou remains the staff of life. Today most Anaktuvuk hunters travel in the wild and lonely mountains on snowmachines or all-terrain vehicles, although a few traditionalists still cling to their beloved sled dogs.

In July 1999 I was invited to the fiftieth-anniversary celebration of the Anaktuvuk Pass post office, where I was given a plaque memorializing my early aerial service to the village.

A DAY IN MY LIFE

After seven or eight years of commercial flying in the arctic, you might say my daily life settled into a routine. On the other hand, it was far from routine—at least from most people's point of view. Here's how I would spend a "typical" day in 1956:

After breakfast at my roadhouse, I walk next door to the CAA building for a pilot briefing. Jim Falls, a good friend, is on duty.

"Where to today, Andy?" he asks.

"Downriver."

"Weather doesn't look so hot," he says, "but you might make it."

I serve an area of about 10,000 square miles, with pinpoints of civilization northwest, east, and south of Bettles Field. My longest scheduled mail run extends 175 miles south, or downriver, where I plan to fly today.

I scan the teletype for weather at Hughes, one of the villages I have mail and freight for. Low ceiling, maybe snow showers, little wind. It looks typical for a fall day and I decide to try the flight. I can turn back if weather worsens.

I return to the roadhouse and pick up a clipboard holding bills of lading, then go to the twelve-year-old plane I am to fly—a

Noorduyn Norseman, a Canadian-built workhorse of an airplane that Wien bought as military surplus. This eight-passenger transport is powered by a 600-horsepower Pratt & Whitney Wasp engine. It weighs 4,240 pounds empty, carries a payload of about 3,000 pounds (fuel, passengers, freight), and cruises at about 140 miles an hour.

The Norseman is a stubby, solid-appearing plane, not especially graceful, but I am fond of it because it flies well, packs a big load—and is dependable. This morning she squats silently, dew heavy on fuselage and wings, waiting patiently for me to bring her to life.

Frank Tobuk meets me at the plane. "Plenty of oil in her, Frank?" I ask.

"Ready to go," he responds. Frank is reliable and I know he checked everything, but I always ask to be sure.

I drain the water traps in the bottoms of the wing fuel tanks and climb up so I can peer into the gas tanks and see the gas level—the gas gauges work, but seeing the gas in the tanks is better. I walk around the plane, checking flaps, ailerons, tail feathers—every movable piece, making sure all bolts are in place and that nothing is likely to come loose in flight.

I turn the big three-blade prop over about a dozen times. This removes oil that may have accumulated in lower cylinders and prevents hydraulicking—damage that can occur to a piston from pooled oil when the engine starts.

I swing into the left seat and fasten my safety belt, peer around to be sure Frank and everyone else is clear of the prop, then prime the engine, pushing and yanking on the noisy priming pump.

"Clear!" I yell, a lifetime habit acquired in the Navy. The habit is so strong I even call this out before starting a plane when I know there isn't another soul within fifty miles.

"Clear!" Frank echoes from where he stands with a fire extinguisher.

I hit the switch to the inertia starter and it whines louder and louder. I hit another switch, engaging the cold engine to the spinning inertia wheel.

The big prop turns sluggishly. It leaps as the engine fires, then makes a half-turn at slow speed, then the exhaust belches blue-white smoke as the engine catches. It misses a few times, then settles to a steady, deep growl. I watch the oil pressure and cylinder head temperature gauges, waiting for one to drop, the other to climb.

I warm the engine thoroughly, which takes about five minutes. While waiting I switch on the radio, call the local CAA, and give Jim Falls my flight plan. There is no wind, so I can take off heading either north or south. I choose to take off to the south, the direction I plan to fly.

I taxi to the north end of the gravel strip where I lock the brakes, hold the control wheel back, and rev the engine, snapping the magneto switch from left to right to both. There is no appreciable change in the rpms, so everything seems normal with the engine. I search the sky for other planes, see none, let up on the brakes, tell Jim on the radio that I am departing, and taxi onto the runway.

All appears well with temperature, voltage, and oil pressure, so I set the flaps and trim tab where they belong for takeoff. Finally, I push and yank the controls, peering out to see if the ailerons are moving as they should, and stretching to peer astern to see if the elevator is responding.

When I shove the throttle forward, all 600 horses roar as the big prop spins and yanks the Norseman down the runway. I have to use a lot of rudder to counteract the torque of the propeller and big engine to keep the plane centered on the runway. Soon the wings take up the weight of the plane and we lift off. I pull her into a moderately steep climb, making sure airspeed remains well above stalling.

174

Stick in hand, I stood in front of the Norseman aircraft during a stop at the Koyukon Indian village of Huslia in 1955. The stick helped me to discourage the unfriendly dog just to my rear. Before Wien Airlines built the village airstrip in 1953, planes always landed on the river—on floats in summer, skis in winter. JIM REARDEN PHOTO

Fragments of low-lying clouds flick past. The engine cowling in front of the windshield vibrates reassuringly as the blunt wings move serenely through the sky. I climb to 1,200 feet and follow the snaking Koyukuk River southward at about 140 miles an hour.

THE BIG BLACK PLANE ROARS over a wilderness of ponds, sloughs, and swampy tundra sprinkled with stunted spruce. A cow moose stands in a tundra lake, her calf nearby. A touch of snow lines the summit of a distant ridge. Occasionally I detour around snow or rain showers. A shiny-coated black bear walks along the bank of a small stream, ignoring the plane.

There are no fences, roads, power lines or billboards. An occasional dogsled trail snakes across the tundra; often these trails stop abruptly on one side of a lake and reappear on the other. Winter ice annually fills the interval. Smoke spikes from the stovepipe of a cabin occupied by an elderly Indian who prefers to live alone. Today he steps out of the cabin and waves. I drop a wing in salute.

As I fly I make minute adjustments in propeller pitch, throttle adjustment, and mixture as the big engine warms to its work. Three big bull moose running single-file pass beneath the plane, their yellow antlers looming large against the dark tundra. Breeding season is near, and I sense their restlessness.

To a stranger, the land I fly over could appear to be as monotonously bland as it did to me when I first flew here. But I have learned the landmarks. Here is a familiar narrow place in the river, there an odd-shaped lake. A few miles farther is a huge rock I've always wondered about. Did a glacier dump it there? I have memorized such landmarks for hundreds of miles around Bettles. When visibility fades, these mind-prints help me to know exactly where I am, where the nearest hills or mountains are. At all costs I want to avoid getting lost or confused and flying into a hill or a mountain.

Thirty-five minutes after takeoff I fly over the ninety-person village of Allakaket, home to a cluster of log cabins and to St. John's in the Wilderness, a lovely two-story log church. This Episcopal mission on the south bank of the Koyukuk has been run since 1923 by Miss Amelia Hill, an Ireland-born nurse, teacher, and lay minister. With her is Miss Bessie Kay, a teacher who arrived in 1932. These two much-loved and respected women have served this village well. Miss Hill conducts services with an Indian interpreter at her side. She is also postmaster, a dentist in emergencies, and occasionally delivers babies.

The Indians of Allakaket live from the land. Fish, game, and furs provide food and income. I fly beyond the village, reducing throttle, adjusting trim, banking, and lining up with the dirt runway. Before the Norseman has touched down, villagers are hurrying toward the runway. The undulating runway causes the hydraulic landing gear to set up a loud clatter as I land. I touch the brakes and roll to a stop at the end. I pull the mixture control. The engine stops, and I snap the magneto switch to off.

"Hi Andy," comes the soft, polite, somewhat shy greeting from each of a dozen or so villagers surrounding the plane. I value the friendship of the Native people along the Koyukuk. I admire their independence, their ability to survive in the harsh climate, their eternal optimism.

After Miss Hill signs for the several sacks of mail, I climb back in the plane and go through the "clear" routine as I start the engine and use brake and power to swing the tail around. I know the prop wash almost knocks some of the people over as I leave. They know it's coming, but stay just the same. There isn't much to entertain in this small village, and my twice-weekly arrival is a break in the routine.

Allakaket lies almost exactly on the Arctic Circle. When I take off and fly south I am no longer properly in the Arctic.

Next stop is the Athapaskan Indian village of Hughes, population sixty-three. Mr. and Mrs. Les James are the white traders. The airplane runway extends into the center of the log-cabin village, and I have barely stopped the engine and opened the door when the schoolteacher steps out of the school, clanging a big handheld bell. Children from all over the village scurry toward the metal-roofed, moss-chinked log schoolhouse.

Nearby, a leisurely working man is banking a newly built log cabin with dirt, anticipating the coming winter's cold. A chorus of howls comes from sled dogs tied near the cabins. A dozen or so men arrive and cluster around the plane, each greeting me individually.

I have flown for about an hour. From one of the barrels I keep at Hughes, I add thirty gallons of aviation gasoline to one of the tanks of the Norseman. The big radial engine gulps a lot of fuel. Lining the inside of the funnel I insert into the tank is a chamois skin that absorbs water and filters sediment from the gasoline. With a low ceiling and unpredictable weather I like to keep my tanks topped off—I never know when I might be forced to fly farther than planned.

With help from several of the villagers I load three 55-gallon drums of fuel oil, rolling them up a plank and through the cargo door. Timmy, an eight-year-old who should have been in school, strays in front of a rolling drum. "Look out, Timmy," I warn. Timmy grins, enjoying the attention, and slowly moves aside.

I lash the drums down—total weight about 1,350 pounds—and soon roar off the Hughes runway.

Next stop is Huslia, another Indian village, population about 150. I skip down the hummocky dirt runway, touching several times before the big plane stays down. I roll the three drums of fuel out of the plane and unload the mail. About twenty-five residents arrive, each saying a polite "Hello Andy." I must search my memory to recall first names of all the people who greet me at these villages. Somehow I always manage to bring their names up.

"The nurse is waiting for you at the other end of the runway, Andy," one of the men reports. I wouldn't have known if he hadn't told me; I can't see the other end of the runway because of a hump midway. About a dozen grinning and chattering residents climb into the plane, and I taxi slowly to the other end of the strip.

"Tickets, please," I kid as everyone jumps out of the plane. They all laugh. "Send us a bill, Andy," one says.

Except for one old truck in the village, some of the residents of Huslia have never seen an automobile.

Eunice Berglund, an itinerant nurse for the Territorial Department of Health and one of my frequent passengers, climbs aboard the Norseman with her medical cases, a down sleeping robe, binoculars, and her 12-gauge double-barrel shotgun. Eunice, originally from Wisconsin and a former Army nurse, loves to hunt grouse and waterfowl.

A young missionary is also waiting with several long, awkward chunks of plasterboard he wants me to fly to Allakaket.

I call "Clear!" and start the engine and swing the plane around. Eunice is still shouting instructions to one of the women who came to bid her goodbye. "I'll see you soon," she shouts over the roar of the Norseman as I head down the undulating runway at full throttle. Huslia is my southernmost stop, so now I head back north.

THE PLANE SLICES THROUGH several spits of early snow. Scattered puffs of fog hang among the spruces here and there. We fly

I put a lot of miles on this Noorduyn Norseman, a Canadian-built workhorse bought by Wien Airlines as military surplus. The eight-passenger transport was powered by a 600-horsepower Pratt & Whitney engine.

JIM REARDEN PHOTO

over a tiny riverbank trapper's cabin; a dogsled stands near the cabin, and grass grows from the cabin's sod roof. A pile of freshly cut firewood lies nearby—a harbinger of the coming winter.

At Hughes again, I pick up freight and mail. Eunice notices I have a bad cold, and I admit my throat is sore and my ears are bothering me. "You're going to get a shot where it counts when we get to Bettles," she promises. I know where she intends to put the shot, and rub my hip thoughtfully.

I stop at Utopia, an Air Force station where I pick up five passengers; four are civilian technicians who have completed a military contract and the fifth is an Indian named Arthur Ambrose who works at a nearby gold mine.

"Strandberg pay," Ambrose tells me when he climbs aboard. Strandberg is the owner of the mine. Ambrose carries a .30-.30 lever-action rifle and a small pack. He wears a red bandana around his neck and peers through dark glasses despite the gloomy weather.

I fly back to Hughes and let Ambrose off. While on the ground I use the radio to call an airborne Wien DC-3 to ask if they can stop at Hughes for the four construction men. They can't, so the men must fly with me to Bettles, where they can catch the southbound DC-3 tomorrow.

I fly north to Allakaket. On the way, a few hundred yards from the Koyukuk River, I spot several people walking on the tundra. I drop altitude and circle, checking to be sure they are OK. They are a long way from the nearest village or cabin. They wave, and each is carrying a bucket. Berry pickers, I decide as I spot their outboard-powered riverboat and climb back to just below the low cloud ceiling and continue on to Allakaket. At Allakaket I land and drop off mail I have picked up at Hughes and Huslia, and the missionary's plasterboard.

As I unload, Eunice calls to one of the men who has come to meet the plane. "Got a moose yet, Joe?"

"Not yet, Miss Berglund," is the rueful answer.

Eunice, in the copilot's seat, opens the door and lifts a mite of a dirty-faced Indian youngster onto her lap and cuddles him while I finish unloading.

Away we roar again, heading for home. Half an hour later as I set up for a landing at Bettles Field, I fly over a huge bull moose with antlers at least four feet across. He stands belly-deep in a slough. As we roar about 400 feet above him, he turns his great antlered head slowly, watching the plane.

Frank Tobuk is waiting to unload mail and freight as I park the Norseman next to the roadhouse. It is 4:30 in the afternoon, and I have flown four hours and twenty-five minutes. In the roadhouse I use the radio to call a Navy plane headed for Anchorage and persuade the pilot to descend from 8,000 feet to pick up the four construction men, who are authorized to travel by military transport. It will speed them on their way home.

I sort mail for Bettles and do paperwork until supper. After supper a Wien C-46 arrives with a load of fuel and freight. Four or five of the roadhouse guests pitch in with Frank Tobuk and me to unload several dozen barrels of fuel, groceries, and freight bound for gold mines I serve.

Except for mealtimes, I have worked straight through from 7 A.M. until 10 P.M. on a more or less typical day in my life as an arctic bush pilot.

THE CRISLERS

One day in mid-April 1953, Herb and Lois Crisler climbed off the Wien mainliner DC-3 with cameras, camp gear, and supplies. Herb, usually called Cris, was a wildlife photographer for Disney Studios. I had not known they were coming to Bettles, nor had we ever met. Cris was a sturdy, bearded fellow, somewhat shorter than his tall, slim wife.

Snow still covered the ground. I was flying a Cessna 170 on skis, but business was slow. Li'l Abner was changing a cylinder on the airplane engine where it was parked near the roadhouse. Nighttime temperatures were still dropping to well below zero. Days were mild, but below freezing temperature.

"Can you fly me to where I can get pictures of migrating caribou?" Cris asked.

In spring the central Arctic caribou herd—half a million animals—migrates north to calving grounds on the north slope of the Brooks Range. In late summer or fall they migrate back to high country.

"Sure, in a couple days. Right now my airplane is being worked on," I replied.

I shared the scene with my Cessna 170 in the Brooks Range in 1953, during a stop at the camp where Herb and Lois Crisler were staying while photographing wildlife. LOIS CRISLER PHOTO

185

The Crislers settled into the roadhouse to wait. After one night, Cris was antsy to get into the mountains. He peered over Li'l Abner's shoulder, eyeballing the various airplane parts scattered around. Apparently he didn't like what he saw. "Andy, can you recommend another pilot who could take us to caribou?" he asked.

"I could, but it would take time for him to get here," I said. "I think I'll be able to go tomorrow." I wanted the job of flying the Crislers.

I went to Li'l Abner. "Can you work straight through until the plane is ready to fly?" I asked. "The Crislers are talking about getting another plane and pilot."

"OK," was all he said. "I'll call you when it's done."

Li'l Abner woke me at midnight. "Plane's ready," he said. "I'm going to bed."

I got up and gave the ski-equipped airplane a flight check of about forty-five minutes, landed and refueled, checked my emergency gear, and loaded several five-gallon cans of extra gas. I called Cris about 3 A.M. "The plane's ready to go anytime you are," I announced.

The snow was frozen and crackled underfoot as Cris and I, carrying parkas and lunch in a paper bag, walked to the warm plane near the roadhouse. Lois would wait at Bettles.

THE SUN WAS WELL ABOVE THE HORIZON as we lifted off, following the John River north. Soon I banked and headed west along the treeless north slope of the Brooks Range. We saw a few small scattered bunches of caribou that appeared to be drifting northwest, but for some time we could find no really large concentration. Finally we saw a long, dark file of about four hundred caribou trekking northwest, heading for a bare, white mountain a few miles away.

During 1953 and 1954, I helped transport Herb and Lois Crisler to and from their Brooks Range camps, where they were filming wildlife. Lois photographed me in 1953 during a visit to one of their camps. LOIS CRISLER PHOTO

"Andy, let me off on that mountain. Then go pick up Lois and our gear and drop her off on this lake," he said, drawing a circle on the map around Tulialek Lake in the northwestern Brooks Range. "You can then pick me up and take me to the lake."

I was astounded by the request. We were a hundred miles or more from Anaktuvuk Pass, the nearest people. I was unfamiliar with the region and even had doubts that I could find the mountain again. Cris had no camping gear, no supplies.

"I can't do it, Cris. Too dangerous," I argued.

He was adamant. "Dammit, Andy, I've got to get pictures of those migrating caribou."

We argued about it for some time before I reluctantly gave in. He pointed out that he had lived in wilderness regions most of his life and had been a coastal hunting guide in Washington state.

I landed with no difficulty, touching down on snow that was wind-packed and several feet deep. I left him all my emergency gear,

187

Herds of caribou may be seen throughout the Brooks Range. JIM REARDEN PHOTO

including a sleeping bag and a plumber's firepot with an extra gallon of fuel. Food included rice and pancake flour. For shelter, I left the covers for my airplane wing and engine.

"Tell Lois to bring enough grub to last until after breakup when you can come on floats to pick us up," he instructed. And then he added: "Andy, if you know anyone who can find a wolf den and get us a litter of wolf pups this spring, we'll pay well for them." He wanted to use the wolves in his filming.

"I'll ask the people at Anaktuvuk Pass," I promised. Then I took off.

I circled for fifteen or twenty minutes, memorizing landmarks so I could find Cris again. It was a clear day and I could see for miles in every direction.

I was amazed at Lois's calm acceptance when I told her I had left Cris alone high on a snowy unnamed mountain with minimal gear. She had utmost faith in his ability to survive. To reassure me, she told me about some of their perilous experiences over the years in their quest for wildlife photos.

Two days later, on the morning of April 21, with Lois in the copilot's seat and the four-place Cessna 170 loaded with food, fuel, tools, clothing, and tents, I flew up the John River bound for Tulialek Lake. We left the last trees behind as I banked and flew west. The aerial map lay on Lois's knees, and she followed our route with a finger.

"That's the way you'll walk out if you have to," I said, pointing to a section of map.

"I know," she answered calmly.

At Tulialek Lake, snow atop the ice was ridged from the pressure of constant howling wind. I couldn't tell how high the ridges were until we landed with a thumpety-thumpety-thump and a loud bang as the skis leaped from ridge to ridge. The snow ridges were hard as rock and nearly a foot high.

The bolt holding the tail ski sheared, and the tail ski lay loose on the snow when I shut down and climbed out of the plane. After unloading Lois and her gear, I reattached the tail ski with wire and pliers and took off to find Cris. As I left, Lois showed no concern but bent to her task of organizing their equipment.

The weather remained clear and I had no trouble finding Cris, who told me he was pleased with the filming of the migrating caribou. I flew him to Tulialek Lake.

These were the first of many flights I made for the Crislers during their year and a half in the Brooks Range making their Alaska wildlife film. Three months after that first flight, Wien pilot John Cross picked the Crislers up at Tulialek Lake.

Wildlife photographer
Herb Crisler packed a big
camera on his shoulder
as I walked alongside my
Cessna 170 in the Brooks
Range. The Crislers
worked in the range
during 1953 and 1954,
getting material for a
Disney wildlife film.

LOIS CRISLER PHOTO

I HAD TOLD THE ESKIMOS of Anaktuvuk Pass about Cris's interest in wolf pups. One of the men there captured five wolf pups at a den and was holding them for the Crislers. I flew Cris to the village, where we found only two of the pups still alive. They resembled dog puppies—tiny, cute, mewling balls of fluff. At Bettles Field, Cris made a plywood pen for them near the roadhouse.

I now flew Cris from Bettles Field to a lake in the central Brooks Range, where he was anxious to continue filming. After leaving him with his camera gear and minimal camping equipment, I took off to return to Bettles to pick up Lois. Water still streamed from my floats and I was slowly climbing when I passed over two huge grizzlies at the lake's edge. Surprised by the plane, both stood on their hind legs snarling, claws ready to defend themselves from the noisy intruder.

Now I had another reason to worry about the Crislers. They didn't have a gun, and grizzlies do occasionally attack people.

A day or so later I flew Lois to the lake, along with the wolf pups and supplies to last the Crislers for much of the summer. I handed my .30-06 rifle to Cris. "You'd better have this. Some of the grizzlies around here aren't friendly."

I worried unceasingly about the isolated Crislers. Although I marked their location on a map, the maps of those days were poor, and there was no certainty that another pilot could locate them. Thus on one of my flights to the Crislers I had pilot Dick Morehead accompany me. After that, two of us knew precisely where the Crislers were. If anything happened to me, Dick could find them.

That fall the Crislers decided to return to Tulialek Lake. It would require several flights for me to move them and their supplies. I arrived for the first flight and landed on the lake. Their tiny mountain tent had been erected on the beach. I could see the Crislers coming toward me, each leading a wolf with a leash. The wolves, no longer squeaking, cute pups, were half-grown. I wondered how they would react to the airplane and to me.

The approaching Crislers disappeared from my sight, and then Lois arrived at the plane without the wolf she had been leading; she had left it with Cris. "Andy," she said, "would you mind crouching inside that tent while we bring the wolves the rest of the way. We don't know how they'll react to a stranger. We'd like to have them close to the plane when you show yourself."

I crawled into the tent, smiling to myself. I had been asked to do many strange things as a bush pilot, but this was a first.

Cris arrived with the male wolf. The animal immediately sensed my presence, and Cris called, "You might as well come out, Andy. He knows you're there."

I crawled out of the tent, and the wolf stared at me, hackles on end. He pulled away from Cris, clearly frightened. We loaded the plane, and Cris got aboard with the wolf. The animal in Cris's arms hid his face from me during the hour-long flight to Tulialek Lake.

I returned to fly Lois and the other wolf to Tulialek Lake, along with enough plywood and other materials for the shelter the Crislers proposed to build and to live in during the coming winter.

On September 3, I made my last flight to the Crislers before fall freeze-up. By then they were living in their plywood "crackerbox," as they called their new home—a shelter that measured eight by ten feet and was six feet high. But it was a decided improvement over a tent: secure, dry, and easy to heat, although crowded.

Three weeks later, the lake was frozen over when I made a drop of mail and frozen food to the Crislers. With the drop I included a scrap of cardboard with a note: "Hope we didn't forget anything. Expect to see you October 15. Wave if everything OK."

As I made another pass over the tiny plywood box, Lois waved. She looked lonely.

A week or so later Disney Studios contacted me. Walt Disney felt it was too hazardous for the Crislers to spend a winter alone at Tulialek Lake and asked me to fly them out. I dropped this message to them October 17, telling them I'd pick them up about November 1.

I landed my ski plane on the frozen lake November 1, and the Crislers started toward the lake with the two wolves on leashes. Lois held back with the smaller female while Cris came to the plane to help load their gear.

With Cris and the male wolf gone, the female wolf became frantic and tried to escape from Lois. It was all Lois could do to hang on and she yelled, "Help, Cris, I can't hold her."

The female, now a powerful animal, was dragging Lois about. Cris called, "I'll be right there." He handed me the male wolf's leash, warning, "Don't turn him loose. We might never catch him again." Then he ran to Lois. Thus, suddenly, I was clutching the leash of a nearly grown, essentially wild wolf.

As Cris left, the animal became frantic, yanking on the leash, struggling to get loose. He bit at the leash. He growled, whimpered, all the time lunging powerfully. It was all I could do to hang on. In his frenzy, twice he clamped down on my wrist with his powerful jaws. Fortunately I wore a heavy jacket and he didn't break the skin; the poor wolf just wanted freedom.

Cris soon returned dragging the other wolf, which he managed to shove into a cage he built. The animal I was holding calmed somewhat and Cris soon retrieved the leash I was holding.

THE CRISLERS WINTERED in Barrow and returned to Tulialek Lake in the spring of 1954. By then the male wolf weighed 110 pounds, the female 90.

Cris had told me he would also like to obtain one or two live wolverines for photographic purposes. I told Cris about a trapper in Fairbanks who had two of these animals. "Let's go get 'em right now," Cris said.

I flew him to Fairbanks. I had expected the wolverines to be young animals, so we were alarmed to find they were two ferocious adults, confined in stout cages, growling threateningly at anyone who approached.

Cris bought them. The cages wouldn't fit in the Cessna, so we loaded them onto a Bettles-bound DC-3 and Cris and I flew back to Bettles Field in the Cessna. We had to handle the cages carefully; the wolverines behaved as if they would tear anyone apart who came near. At Bettles we loaded the two cages into the Norseman.

"Cris, I'll fly those critters to the lake for you," I told him, "but if one of them gets loose while we're airborne, I'm going to shoot it."

"I'll beat you to it," he said. "They aren't very friendly, are they?"

I landed the Norseman on the ice of Tulialek Lake just twenty-four hours after I told Cris about the wolverines. We unloaded the cages, and I was glad to see the last of those savage critters.

I continued to fly supplies to the Crislers, making drops when the ice was too rotten to land, and before open water. After breakup I landed at the lake on floats.

During the spring of 1954 Cris found a wolf den not far from the lake and took five pups from it. He and Lois raised the five, with help from the one-year-old wolves they already had. Their wolves ran free, essentially wild, and became an important element in the Crislers' film.

I made several trips to the Crislers' camp that summer. By September 26 the lake was frozen over, but the ice wasn't thick enough for safe landing. I could only drop supplies to them.

Cris spelled out a message for me with five-gallon fuel cans. "Oct. 10 if ice," it said, meaning he wanted me to pick them up on that date if the ice was suitable for a landing.

I arrived over their camp with the Norseman on October 10 to see a message spelled out, again with the five-gallon fuel cans: "Ice 13," meaning the ice on the lake was 13 inches thick. I circled and landed on clear glare ice so slippery I had little directional control and couldn't taxi to the Crislers' dunnage piled on the shore. I shut the engine down and waited for Cris to arrive.

When he arrived I had him loop a rope around the tail and swing the plane to face the dunnage pile, fired the engine up, and taxied up beside the pile. Without the guiding rope the plane was uncontrollable on the glassy ice.

The first two wolves raised by the Crislers had adapted to life in the wilds and were left on their own in the mountains. The five five-month old pups could not have survived, so the Crislers loaded them into the plane. They behaved much like very shy dogs and gave no trouble.

By late October 1954, the Crislers with their five wolf pups had returned to their cabin home in the Tarryall Mountains of Colorado. Lois Crisler's book *Arctic Wild,* describing their arctic Alaska experiences, was published in 1958 by Harper Brothers. The fine Disney film resulting from their work was also titled *Arctic Wild.*

CHAPTER FOURTEEN

LOST PILOTS, BROKEN AIRPLANES, AND OTHER TRAGEDIES

Clarence Rhode, Alaska regional director of the U.S. Fish and Wildlife Service, was at the controls of the Grumman Goose N720 as it sped down the runway at Fairbanks International Airport at 12:10 P.M. on August 20, 1958, on a routine flight into arctic Alaska.

With Rhode were his 22-year-old son Jack and Fish and Wildlife agent Stanley S. Fredericksen. Aboard were two hundred gallons of aviation gasoline to cache for later patrols in the arctic planned by Fredericksen.

They were to fly to Porcupine Lake on the north slope of the eastern Brooks Range. Rhode radioed the Fairbanks office of the Fish and Wildlife Service when he was over the Middle Fork of the Chandalar River. Then, slightly more than two hours after leaving Fairbanks, he reported that N720 was on the water at Porcupine Lake.

No one ever again heard the radio of Grumman N720. Part of the reason may have been severe atmospheric conditions that for many days blacked out radio signals in northern Alaska. With

197

signals blacked out, Rhode could no longer report his position. No one back in Fairbanks could know where he intended to go from Porcupine Lake.

That same day, Rhode flew the Grumman to Peters and Schrader Lakes, about forty miles northwest of Porcupine Lake, where the wife of a hunting guide saw them and where they visited an International Geophysical Year party camped at Peters Lake.

They told the IGY party they planned to fly west and possibly return to Porcupine Lake. These people later reported they had seen the Grumman leave Schrader Lake and head west about 5 P.M. that day. This was the last seen of the plane. Grumman Goose N720 and its three occupants had disappeared.

Weather in the Brooks Range was generally bad that August. Frequent storms and high winds hampered the aerial search for N720, which began a few days after it disappeared. Aircraft of the Fish and Wildlife Service, the Bureau of Land Management, the Civil Aeronautics Administration, the Air Force, and private pilots flew in the search. Planes flew day and night; night flyers watched for campfires.

For a time, virtually all the searchers flew from Bettles Field and lived at my roadhouse. One day I counted forty guests and more than twenty visiting planes, almost all involved in the search. We were crowded, with a lot of sleeping bags scattered around our guest rooms and lobby, but no one complained.

Rhode's plane had survival gear aboard, and all three men were expert woodsmen, giving searchers hope that wherever the Grumman was down, survivors could be camping. Searchers believed the men could last for months, even through the winter.

The search for Rhode's plane was the most extensive and longest ever made in Alaska up until that time. More than 260 people and as many as twenty-eight planes were in the air at one time. The search covered more than 300,000 linear miles over an area of more than 300,000 square miles—about twice the area of the state of California.

Many heavily wooded areas that would camouflage a downed plane exist south of the Brooks Range, a land of canyons, rugged mountains, and deep lakes. It is a vast land where a missing airplane is hardly a flyspeck. The thousands of square miles of uninhabited land I flew over during my years at Bettles Field hid many secrets, including the locations of lost pilots and planes in addition to the Rhode plane. This was a fact I flew with as my planes coursed the tremendous distances between human habitations.

Speculation about what might have happened to the Grumman ran wild. Did the load of gasoline explode in flight? Had Rhode landed the amphibian on a lake with the wheels down, wrecking and sinking the plane? Had he somehow flown over the nearby Arctic Ocean, crashing there, never to be found? Had the plane plunged into a dense spruce forest where it couldn't be seen from the air?

After the search dragged on for a couple of weeks, the State Department approached Russian officials, asking if Russia had shot down a plane answering the description of the Goose. With the gasoline in its tanks and extra fuel on board, the plane had a potential range of about two thousand miles. This was really grasping at straws.

During the search, an unidentified emergency radio signal raised hopes, as did a flare reportedly seen near the western end of the Brooks Range.

No pilot ever located the plane. From the time of its disappearance until I left Bettles Field in 1965, I watched for the Rhode plane whenever I flew; I probably flew over the wreckage without seeing it.

In late August 1979, almost twenty-one years to the day from when the plane disappeared, two adventurous women hikers crossing a 5,700-foot-high pass at the head of the Ivishak River found what was left of Grumman N720. The site was only twenty-five miles southwest of Porcupine Lake.

It is likely that snow covered the wreckage much of the year, making it virtually impossible to see from the air. The plane had slammed into the mountain at high speed, tearing it apart. Pieces were scattered across a steep talus slope. Death for the three occupants must have been instantaneous.

This crash haunted me for years. I was retired on my Pennsylvania farm when Grumman N720 was finally found. Though I was no longer involved in Alaska flying, closing the search brought some peace of mind. When there is no ending to a tragedy, thoughts about it linger.

BIG-GAME GUIDE BUD BRANHAM, one of the best of the best of pilots and guides, got himself into big trouble in the Brooks Range in 1953. Even the best can make mistakes.

Bud usually guided his trophy hunters from his Rainy Pass Lodge in the Alaska Range. However, he had found a fine location for hunting Dall sheep at Loon Lake, about eighty miles northwest of Bettles. Mountains surrounding the lake poke up to 6,000 and 7,000 feet, but the lake can be approached either from the north or the south through a pass.

Bud hunted this area for a number of years, flying a twin-engine Grumman Widgeon amphibian. Built for the military in the 1940s, these rugged planes were fitted with two 200-horsepower Ranger in-line engines. Performance was poor. The plane was underpowered. Later Widgeons were repowered with better engines, but Bud's plane had the low-powered Rangers.

In the fall of 1953, Bud flew to Loon Lake before sheep season to set up his camp. He then flew to Fairbanks to pick up two nonresident hunters and headed back with them to Loon Lake. Before landing, he decided to fly to the back side of a mountain to show his clients a band of sheep.

He flew into a canyon that climbed abruptly and was too narrow for him to turn around. The sheep were low on the mountain as Bud flew near. He had misjudged the performance of his plane. Using full power, he climbed, but didn't gain enough to clear the ridge. A flat spot appeared ahead, and Bud, realizing it was his only out, crash-landed the plane. One of the hunters sustained a severe back injury, the other suffered minor injuries. A hole was punched in Bud's chin; when he drank water, it leaked onto his lap. The Widgeon was totaled.

Bud was tough. Although injured, he managed to use the plane's radio to call Bettles CAA and report the accident.

My floatplane was in Fairbanks for an inspection, but a close friend staying at the lodge owned a nice floatplane, which he loaned me. When I reached the area and saw the situation, I radioed the Air Force's Tenth Rescue Squadron and asked them to send a helicopter. The chopper arrived after dark, but with spotlights the pilot managed a night landing and picked up the two hunters. In three hours he flew them to the Fairbanks hospital.

Despite his injuries, Bud made his way to his camp on foot, where he remained overnight. The following morning I flew him, his assistant guide, and his camp cook to Bettles, where they caught the Wien DC-3 mainliner to Fairbanks for connection to their homes. Their hunting season was ended.

Remains of Bud Branham's Widgeon were retrieved in the early 1990s through auspices of the Alaska Aviation Heritage Museum, in

Anchorage. Repairing it was beyond the capabilities of the museum, but new owners in California were rebuilding the plane.

AN ARCTIC AVIATION TRAGEDY befell one of my best friends, Jim Falls, a CAA controller at Bettles Field, who came to Alaska from Minnesota. He lived to fly and to hunt.

Jim owned a Piper Super Cub, the two-place, tandem-seat airplane favored by many Alaska guides and hunters because it can get in and out of small places. This 150-horsepower plane is easy-to-fly, dependable, and rugged. Jim flew almost as many hours a year as I flew on my scheduled mail runs. He knew the best places to fish, where to find moose, caribou, sheep, and ptarmigan. He lived largely on the wild game and fish he shot and caught.

Wanda Falls, Jim's sister, visited Jim at Bettles Field and liked it so well that she decided to remain. Eventually she married Robbie Robbins, the CAA station manager, and was quite at home in our little community.

Tom Delock, Jim's twenty-year-old nephew, Wanda's son from a previous marriage, also visited Jim and adopted Bettles as his home. He often flew as gunner with Jim on aerial wolf-hunting forays.

During the 1950s, aerial hunting of wolves was popular in many parts of Alaska, and the Koyukuk Valley attracted many of these airborne hunters because of the overabundance of wolves. The Territory paid a fifty-dollar bounty for each wolf; the wolf skins, prized for making parka ruffs, brought an additional thirty-five to forty dollars. In spring when there was plentiful daylight, snow was deep, and flying conditions were good, Jim and Tom sometimes killed as many as one hundred wolves a month.

I returned to Bettles late one day from a mail run to learn that

Jim and Tom hadn't returned from an aerial hunt. Jim had planned to be back by 2 P.M.

It was too late in the day to initiate a search, so everyone at Bettles Field went to bed with hopes that the men's plane had given trouble and they had safely landed somewhere. Both Jim and his nephew were outstanding woodsmen, capable of taking care of themselves with the emergency gear in their plane.

When they didn't show up the next day, the Fairbanks Civil Air Patrol and the Tenth Rescue Squadron of the Air Force started a search. Day after day the planes came back to Bettles without finding a trace of the missing Cub. After two weeks, the search was discontinued.

The Falls family in Minnesota appealed to their Congressman, who pushed the right buttons to resume the search. No sign of the plane was found on the first day of the renewed search. On the second day the wrecked plane was located. Both men had died on impact. Evidence indicated Jim was chasing wolves along the frozen Koyukuk River just north of Bettles Field and, in the excitement, apparently forgetting about the surrounding terrain, flew head-on into an outcropping.

It was a terrible loss and a tragedy that saddened everyone in the close-knit community of Bettles Field.

ONE NEVER KNOWS how the search for a missing plane will end. Every Alaskan pilot follows such searches carefully, and many pilots volunteer their planes and their time to help.

One search was that for Jules Thibedeau, who I regarded as a hard-luck pilot. I knew Jules only casually; I heard other pilots joke that Jules walked more miles away from airplane wrecks than he had flown.

I made thousands of takeoffs and landings from this section of the Koyukuk River at Bettles during my seventeen years as an arctic bush pilot. Here I'm flying a Cessna 180.

JIM REARDEN PHOTO

When I knew him, Jules lived at Barrow and flew from there. He was constantly purchasing worn-out airplanes, probably one reason he kept crashing. He seemed to be broke or near-broke about every time I ran into him.

I was doing paperwork at my desk in the roadhouse one Sunday morning when I glanced out the window and saw a strange Taylorcraft tied down in front of the FAA building. It was bitterly cold and windy. Soon someone arrived at the aircraft, and as I glanced up from time to time, I saw he was firepotting the plane with a small heater, trying to warm the engine so it would start. Due to the strong wind he wasn't having much success.

I sauntered out to the airplane. The T-Craft, similar to the one I owned in 1948, had no starter or generator, requiring it to be hand-propped to start.

"Can I help?" I asked

Jules, an independent soul, answered, "Nope, I'm doing all right."

Obviously he wasn't. I offered him the use of my big Herman Nelson heater.

"I don't have any money to pay for the fuel for it," Jules said. "I spent my last dime buying this airplane."

"I'll give you the gas," I offered. Clearly he would never get his airplane started with his little firepot. After he agreed to accept the loan of the Herman Nelson heater and the free gas for it, I pitched in to help.

As we worked, I learned he was en route to Barrow after buying the T-Craft. We fired up my big heater and started to heat the engine with it. Then we noticed a gas drip in the vicinity of the carburetor and had to shut the heater down, fearing a fire. We found that the fuel line from the filter to the carburetor was barely in place, held by only about two threads.

Since no proper gas lines were available at Bettles Field, Russell "Mac" McConnell, a mechanic for the FAA at Bettles, built him a temporary fuel line out of copper. This didn't meet FAA requirements, but since there was no suitable material available, it would have to do.

"I'll get a proper fuel line made up for it when I get to Barrow," Jules promised.

With the new fuel line in place, we again started my heater and soon the airplane engine was warm enough to start. I volunteered to hand prop it while Jules sat in the plane and monitored the controls.

When I pulled the prop through I was amazed to find virtually no compression. I doubted the airplane would start. I could yank the prop and it continued to spin after I released it. But somehow the engine did start and Jules sat in the airplane, allowing it to run for a while.

It was late in the day, and since it is a good six hours of flying time to Barrow from Bettles with a T-Craft, I said he could spend the night at the roadhouse at no cost. Weather permitting he could leave in the morning.

Jules and I checked the fuel in the plane and I asked him why the wing tank was empty. "There's a big block of ice in it," he said. "I haven't had time to thaw it out. I can hear it clunk whenever I raise or lower the wing."

I could scarcely believe what he said, but I let it go. It was his airplane, his life.

Jules shut the engine down and spent the night at the lodge. I told Canuck and Frank Tobuk that if they felt like it, they could help Jules start his engine the next morning and see that he got away.

My Monday was busy. I left Bettles early to meet a mainliner Wien flight at Hughes. From there I distributed incoming freight, passengers, and mail. A big high-pressure ridge was sitting on the north slope of the Brooks Range, pushing strong winds through all the passes. I doubted anyone in a light plane would negotiate the fearful turbulence of Anaktuvuk Pass on such a day, and I assumed Jules would postpone his flight to Barrow.

At day's end, when I arrived back at Bettles Field, I learned he had left Bettles Field shortly after I had, en route to Barrow.

That evening we received an inquiry from Barrow. "Have you seen Jules Thibedeau's airplane?"

Wien had radio operators stationed at Umiat, a former Navy runway on the north slope of the Brooks Range between Bettles Field and Barrow. We called Umiat and asked about Thibedeau. An airplane fitting the description of his T-Craft had landed at Umiat about one o'clock in the afternoon. The pilot had filled his nose tank with gasoline he had with him in the plane and departed about 1:30. The temperature at Umiat had been between 65 and 70 below zero all day.

The Air Force search and rescue squadron from Fairbanks launched a search for Jules the next day, concentrating on the 170 miles between Umiat and Barrow. Searchers combed the area for about two weeks without finding a trace of the plane.

"We'll search one more day," the Air Force announced.

On that last day, one of the Air Force planes flew along the shoreline east of Point Barrow as far as the mouth of the Colville River. As the plane turned, preparing to fly back toward Point Barrow, an observer spotted the T-Craft, down on the Arctic Ocean ice pack. Jules was there, alive.

A ski-equipped Wien Airlines light plane picked Jules up and returned him to the Barrow hospital, where he eventually recovered from frozen hands and feet. Jules was still flying when I left Alaska, but some years later I heard he had died in a plane crash.

I NEVER BECAME ACCUSTOMED to the tragedies I encountered as a bush pilot. I often flew bodies, and I helped with searches for lost planes and people. I flew passengers to and from many funerals.

I often flew personnel connected with the Arctic Research Laboratory at Barrow. One such party that I flew to a work site was from Yale University and was headed by Professor Jack Campbell. Don MacVicar, who was an associate of Campbell's, and a student were conducting archaeological studies around Chandler Lake, where digs showed many signs of habitation by early-day people. At the summit of the Brooks Range, Chandler Lake is about fifteen miles long and two to three miles wide, with a depth of forty to sixty feet. When the wind blows through Chandler Pass, which is often, huge waves roll down on the lake.

MacVicar and the student were camped on the lakeshore. One morning they used a twelve-foot folding canvas boat powered with a small outboard to cross the lake to their work site. The wind came up and became so strong it interfered with their work. They decided to return to camp. The lake was rough when they started out, but they decided to chance a crossing. Each wore a Navy-type life jacket which required inflating—the user blew into a stem.

As they reached the middle of the lake, the wind pushed the boat's bow downwind. At this, waves washed over the stern and swamped the boat and sank it. The student, without trying to inflate his life jacket, swam for the shore. The student's last view of MacVicar was of him trying to inflate his life jacket. Evidently he was unsuccessful, for he disappeared.

The search for his body continued for more than a month, including sweeps by Navy divers I flew to the lake. Late in the search, depth charges were set off in hopes of bringing the body to the surface, to no avail. MacVicar's body was never found.

The Arctic Research Laboratory flew his parents to the lake, where they conducted a touching memorial service. I will always carry with me the image of those sad parents standing with bowed heads on the shore of that wild arctic lake that had claimed the life of their son.

NEVER A DULL MOMENT

One cold wintry night a stranger arrived at Bettles Field on the Wien DC-3 mainliner and came into the roadhouse. "I need a place to stay until I can arrange to be flown to a couple of claims I want to prospect," he told me.

After living in the roadhouse for about a week, the stranger started supplying whiskey to the young Air Force men who operated the TACAN long range navigation aid. I allowed no drinking in the roadhouse and told the man his behavior was unacceptable. I suggested he move out.

The only other shelter in the area was at Jim and Verree Crouder's home and store five miles away, at Bettles Village, which occasionally accommodated a boarder. The stranger moved there. Problem solved. We forgot about him, but only for a time.

A couple of weeks later, Jim and Verree had to fly to Fairbanks on business. Since bush Alaska residents are very trusting, Jim asked the boarder if he would watch the store for a couple of days. The boarder agreed and then convinced Crouder he was a pilot.

"I want you to buy me an airplane at Fairbanks," he requested. Crouder agreed to at least price airplanes; the stranger wrote out a $25,000 check for the proposed purchase.

A few hours after Crouder was out of sight on his way to Fairbanks, the boarder walked to Bettles Field. "I've got to fly Outside [to the states]," he told Jim Falls. "My family is ill and it's an emergency. Will you fly me to Fairbanks so I can catch a plane home?" He was convincing, and Falls agreed to the flight.

Falls flew the man to Fairbanks. During the flight, for some reason Falls grew suspicious. At Fairbanks the stranger acquired a car and started south on the Alaska Highway. Falls found Crouder in Fairbanks and told him that his boarder had flown the coop, and his behavior seemed suspicious.

Crouder asked the Territorial Troopers to detain the man. They said they had no reason to arrest him, but they suggested that Crouder make a citizen's arrest.

The police agreed to set up a roadblock on the Alaska Highway, intercepting and holding the fleeing boarder until Crouder could drive the twenty or thirty miles to arrest him. Among items found in the man's suitcases were many of the Crouders' possessions. The crook had scooped up everything of major value in the store and home he could stuff into suitcases.

When the Crouders returned home they found their store and home thoroughly rifled. A store safe was blown open. Among personal effects left by their late border were a dozen or more business cards bearing differing names—apparently all aliases. Later we learned the man was a dangerous escapee from the Arizona state prison, a convict who had shot a state trooper in New Mexico.

The $25,000 check he had written was, of course, no good.

ED PAULSON, WELL INTO HIS 80s, born in Sweden, a longtime resident of the Bettles area, had mined and prospected the region most of his life. In his declining years he elected to remain in the Arctic despite the long, dark, cold winters.

Ed built himself a 10-by-12-foot log cabin on the far side of the Koyukuk River from Bettles Field. He was alone, and didn't even own a dog. Most old-timers who lived alone owned one or more dogs.

Over several winters, on my frequent flights I noticed that each winter day, Ed left his cabin and cut down three small, dead spruce trees for firewood. I wondered about that and was concerned for him, feeling he should have a stockpile of firewood.

One winter day I walked across the frozen river to Ed's cabin and knocked on the door. He invited me in and I found a place to sit in the tiny, neat cabin while he poured us each a cup of coffee.

"Ed," I said, trying to word my offer carefully, for I knew he was determinedly independent and I didn't want to offend, "I'd like to help you with your firewood. I notice you cut only three little trees every day. I have a chain saw, and I'd like to help you pile up enough firewood to last you for a couple of months."

He stared at me for a moment, then tears came to his eyes.

"Yah, Andy, dat's awfully nice of you," he replied in his strong accent, "but, no, I'd rather you didn't help me."

He was silent for a moment and I was afraid I had offended him. Then he explained. "Andy, winter's an awfully long and cold time. I could get very lazy if I didn't have something to do. I never cut more than a day's firewood at a time because I know that on the following day I have to get off my lazy butt and go out into that cold and cut another day's supply. It not only occupies my mind, it gives me the exercise I need."

I had never considered it from that standpoint.

For several more winters I watched Ed continue with his self-imposed daily chore. Eventually it became obvious that cutting even

Drums full of oil were among the endless variety of items we carried by air. Here I'm getting some help loading drums into a Norseman bush plane.

JIM REARDEN PHOTO

three small trees a day was almost more than he could handle. At that point I convinced him to enter the Pioneers' Home at Fairbanks, where he no longer had the daily chore of cutting firewood.

ONE DAY I RECEIVED A CALL from the Alaska Native Service hospital at Tanana. The bureaucracy had become confused: A six-year-old Native girl, recently discharged from the hospital, had mistakenly been flown to the village of Huslia. She should have been flown to her home at Koyukuk, a village near the confluence of the Yukon and Koyukuk Rivers. Would I pick her up and fly her home?

Shaking my head over such a silly mistake, I flew my floatplane to Huslia despite the lousy weather, picked her up, and flew her the seventy miles to her village. She had no sooner climbed out of the airplane when Dominic Vernetti, who owned and ran the local trading post, rushed to my airplane. He was very upset.

"Andy, my mother-in-law is terribly ill. Please fly her to the hospital," he requested.

The weather was still lousy, and the Alaska Native Service hospital at Tanana was 150 miles away. But I never turned down a medical emergency if there was any hope I could get through. I refueled the plane and removed a seat so a pallet could be made up for the elderly patient. She was too ill to sit up.

"Thanks, Andy," she managed to say as we tried to make her comfortable.

The flight to the hospital was a nightmare. I dodged heavy rain showers and fog patches, and the plane shuddered frequently in heavy turbulence. I worried about the sick woman, but she said nothing. I finally landed on the Yukon River in front of the hospital at Tanana, tied the plane, and trotted into the hospital to look for help. The first person I found was the doctor.

"I've got an awfully sick woman in the plane," I told him. "Can we get some help to carry her into the hospital?"

"Let me examine her first, Andy," he said.

We returned to the plane and I waited on the bank in driving rain while he stood on the airplane float, opened the door, and bent over the woman with a stethoscope. He took a long time about it and repeatedly moved the instrument. Finally he looked at me and shook his head.

"She's gone. You might as well fly her home."

I was shocked. The poor woman had died en route. I wondered if the turbulent flight had brought about her death and asked the doctor about it.

"I doubt it, Andy. It's impossible to tell now, of course, but chances are she would have died whether she was in the plane or otherwise."

I had already flown eight hours that day and darkness was near. The weather was still terrible.

"Doc, I'm worn out. I just can't fight the weather all the way back to Koyukuk today. With the weather as it is I'll be lucky to make it home to Bettles."

I persuaded the doctor to accept the body and to send it back to Koyukuk on the mail plane—a larger plane—next day.

My return flight to Bettles that day was through the gloomiest of weather. I encountered fog and heavy rain and fought strong winds. I was filled with sadness for the poor woman, and it was all I could do to concentrate on my flying.

I ONCE TOOK OFF, alone in the plane, with a load of freight and one very large sled dog that I was to deliver to a village. As the plane left the ground, the dog became frightened and developed an intense longing for security. He wedged himself between me and the controls, paws on my shoulders, furry head against my cheek.

The plane was all over the sky as I wrestled with the frightened animal. I suppose it took only about thirty seconds to work him off my lap and shove him behind my seat, but it seemed forever. After that, I always tied a dog before flying with it.

ONE TERRIBLY COLD WINTER DAY I had an unusually heavy load in the Norseman and asked Canuck to ride to the south end of the field with me to help get the plane turned. At the end of the field he jumped out to push the tail around. The snow was sticky and I had trouble swinging the plane, even with Canuck pushing.

I blasted it. The prop wash from that three-bladed prop swung by the 600-horsepower engine knocked Canuck down. I had lost sight of him and was unaware of this and kept swinging the plane, and the tail swung over him as he lay on the ground, and the flying bar under the tail hit his nose. It was close to 50 below, and his nose was numb to begin with.

I started to head down the runway for takeoff when I saw Canuck next to the runway, holding his nose with both hands. He was outrunning the airplane.

I shut the plane down, leaped out, and ran and caught him. "Canuck, what's the matter?"

"Cut my nose off," he mumbled, still holding his hand over his nose. He was in pain, with tears in his eyes.

That scared me. "Let me see," I urged. He shook his head, closely holding his nose. "Come on, let me see it." He reluctantly let me pull his hands away from his face. I was relieved to see his nose still in place.

"Your nose is still there," I assured him.

His nose was cold and numb. The hard bump from the plane had added to his misery, and he really thought he had lost it. I didn't dare laugh.

EUNICE BERGLUND, THE TERRITORIAL Public Health nurse, was an angel with a heart of gold. She could also be forceful when the occasion demanded.

I once called the Public Health office at Fairbanks and asked for Eunice when I had what I thought was a strep throat. It was Eunice's day off, so I asked the nurse who answered the phone how to best treat my throat. We completed our conversation and I went about my business.

That day Wien made an extra DC-3 mainliner flight to Bettles with freight and mail that had accumulated in Fairbanks. I was standing on the tarmac when the DC-3 wheeled to a stop and the passenger cabin door opened. Out stepped Eunice.

"What in the world are you doing here, Eunice?" I asked in surprise.

"I'm here to treat that strep throat of yours, and I'm going to teach you not to call me on my day off," she snarled. "Better prepare yourself. I've got the biggest needles I could find with me, just for you, Andy."

Over the next couple of days she delivered as promised, giving me three shots in my rear with the biggest needles I've ever seen on a syringe. My strep throat disappeared.

That was Eunice. She never ignored a medical need she considered serious. Her special trip to Bettles Field to treat my strep throat was typical. She was also having a little fun with her favorite pilot.

A 1954 outbreak of polio in Interior Alaska brought Eunice and her assistant to Bettles early one morning to combat the disease among the Koyukuk villagers. She wanted to visit every village in her sector to administer the newly developed polio vaccine. I took off with the two of them within a few minutes of their arrival and flew all day, all night, and well into the following day, from village to village until she had done everything possible to get the vaccine into susceptible villagers.

The vaccine proved effective. There were many cases of polio in the Fairbanks area that fall and winter, but not one was diagnosed in Koyukuk country.

I BECAME VERY ATTACHED to Sammy and Ludy Hope, an elderly Native couple and the only year-round residents of the village of Coldfoot. In 1900, greenhorn gold seekers had gotten as far up the Koyukuk River as this location, then lost their spirit, turned around, and left. In other words, they got cold feet; thus the name of the place.

Sammy made a good living trapping the fine arctic furs of the region and worked for various gold mines in summer. Ludy was a full-time housewife. When they needed supplies they mushed their two dogs the eleven miles to the trading post at Wiseman.

Sammy, about five feet tall, was thin as a rail and weighed perhaps 120 pounds soaking wet. Ludy stood a full six feet and weighed approximately 300 pounds. She wasn't fat; she was simply big. She always had a smile, and in her eyes Sammy was the world's greatest man.

Joe Ulen, at Wiseman, operating the local Alaska Communications System radio, called Fairbanks each evening to relay messages or orders that accumulated during the day. Residents of the region commonly eavesdropped on his frequency to pick up whatever information they could from the personal messages he sent. Joe's radio signals were often disrupted by atmospheric conditions, but he blamed this on eavesdroppers who were "stealing my signal strength."

It was amusing to hear Joe announce, "OK, all you people listening in, turn your radios off. Sammy and Ludy, you two at Coldfoot, you're stealing my signal strength. Turn your radio off until I get these messages through."

Ludy enjoyed her vodka and occasionally approached me with, "Mr. Andy, when you go to Fairbanks, would you get me two bottles of vodka. None of that weak stuff. I want 200 proof." It was forbidden for a pilot to sell or deliver spirits to the Natives, but for my good friends Ludy and Sammy, the chance was worth the reprimand, and I occasionally helped them out.

I recall an incident when Sammy and Ludy drove their dogs into Wiseman to catch my scheduled mail flight. They planned to stop at Bettles Field and ride my next downriver flight to Hughes. It didn't work out quite that way.

With the crowded four-place airplanes of the time, and with Ludy's 300 pounds, I had to remove the rear seats to accommodate her. After I had pushed and pulled for about five minutes in an unsuccessful attempt to stuff her into the plane, I gave up.

"Sammy, she's your wife. You get her into the airplane."

Sammy bent over at the door of the plane and had Ludy sit on his back. He then raised up, rolling her into the plane. It took a few minutes for Ludy to squirm around, brush herself off, get seated, and regain her composure. Then she was ready for the flight, which I made nonstop to Hughes; I wasn't up to the challenge of having to unload and load Ludy again at Bettles.

One time, Sammy stopped at Bettles for a visit. We were in my

Planks running from the ground into the plane created a ramp for rolling oil drums aboard the Norseman.

living room, and Sammy had had a few drinks and was feeling pretty good. I was trying to get some black coffee into him before flying him home to Coldfoot. He noticed a bowl of plastic fruit, including grapes. Suddenly he began trying to pull the plastic grapes off the plastic vines.

I knew he might be hungry, but he certainly couldn't have been that hungry. "Sammy, you can't eat those grapes," I warned, trying to let him know the fruit was phony.

Sammy became indignant. He sat up as straight as he could and looked me in the eye. "And why not? You eat when you come to my house don't you?"

Ludy died of natural causes. Not long after that, Sammy became seriously ill and I picked him up at Coldfoot and flew with him to Bettles. I stopped to refuel and flew on, trying to get him to the hospital at Tanana. He died en route.

I've always thought that Sammy lost his will to live after he lost Ludy. Both were well up in years when they left this world, and they had had long, happy, sometimes hilarious lives.

FRED AND JOANNE PITTS mined on Lake Creek, a tributary of Big Lake, thirty miles east of Wiseman. One day, late in August, Joanne and her two-year-old son were walking near their cabin as their two pet malamutes ran ahead. Suddenly Joanne heard the dogs screaming. She ran around a bend in the trail to see two wolves attacking one of the dogs. Several wolf pups milled around the action.

Joanne grabbed her son and fled home. Shortly afterward, the surviving dog joined her. All that was left of the other dog was a piece of tailbone; the wolves had eaten or carried off the entire animal.

The next day I flew in with mail for Fred and Joanne. As I

neared landing, I saw a white wolf on a nearby trail but paid little attention, because wolves were common. Joanne was packing a rifle when she arrived to pick up the mail. She told me about the attack.

"I'll run that white wolf off when I leave," I promised.

The wolf was still nearby as I departed. I buzzed it several times, making sure it fled. When I last saw it, that wolf was hot-footing it across the hills, still peering over his shoulder. The big silver bird and the roar of a 225-horsepower engine with the prop in flat pitch ten feet overhead had obviously impressed him. I never again saw a wolf in that particular area.

THE OLD SAW THAT "flying consists of days of boredom punctuated by seconds of sheer terror" hits close to the truth. In 1954, three times within a month, airplane engines quit on me in flight.

When an airborne airplane engine suddenly stops, it's panic city. The plane is going to land immediately. Where and how it lands spells the difference between life and death. In the 32,000 hours I flew small planes from Bettles Field, I experienced a dead engine in flight only five times. Yet strangely, three of those failures came during that brief period in 1954.

That year the U.S. Coast and Geodetic Survey was mapping northern Alaska. This required a lot of flying in small planes and helicopters. Wien didn't bid on the work, and the pilots who contracted to do the flying weren't able to keep up with the work. So the agency chartered me and my planes to help out.

My first job was to move forty people and about thirty tons of equipment from Stevens Village on the Yukon River to Umiat on the north slope of the Brooks Range. I arranged to fly personnel and equipment from Stevens Village to Bettles Field with the Norseman. From

Bettles the haul to Umiat was completed by a Wien C-46 transport.

Since it was summer with twenty-four hours of daylight, I could fly as many hours as I could stand. On the day I started, I completed my mail run, then flew to Stevens Village and ferried men and equipment to Bettles Field with the Norseman on wheels.

On my third trip, loaded to the maximum, I approached Bettles for a landing and reduced throttle. The engine promptly quit. The sudden silence after the roar of the Norseman's 600-horsepower engine was startling. As I shoved the nose down to keep flying speed, the whistle of air over the wings and across the fuselage was the only sound I heard. Fortunately I was within gliding distance of the runway and made a dead-stick landing.

Canuck found the nine studs holding the carburetor to the manifold had apparently been stretched from an engine backfire. There was about a quarter-inch gap between carburetor and manifold. Thus when I reduced power the mixture was too lean and the engine stopped. Canuck tightened the nuts, and two hours after my dead-stick landing I was again in the air—same plane, same mission.

After the C-46 hauled the men and equipment to Umiat, the Coast and Geodetic Survey chartered my Cessna 180 floatplane to distribute the men and equipment to various campsites on the North Slope. On this job I logged twelve to fourteen hours of flight time a day. At Umiat I took off as soon as the morning fog lifted and quit flying only when the sea fog rolled in from the north, which was usually around midnight.

All went well until one of my takeoffs from the lake at Umiat. I had climbed to about a thousand feet when, bang, the engine quit. Again all I heard was the whistle of wind over the wings and fuselage. Luck was with me. There are many lakes in the area, and I was able to land on one of them dead stick, straight ahead.

The engine was shot. Wien flew a new engine and two mechanics to Umiat and the change was made where the plane landed.

I finished flying for the Coast and Geodetic Survey with a sigh of relief; two engine-out experiences within a week or so was a little much. If either engine loss had occurred at the wrong place, it could easily have ended my flying career—maybe even my life.

But I wasn't through with losing engines for that season. A few weeks later I loaded a Cessna 180 equipped with a ski-wheel combination and took off from Bettles Field on a regular mail run to the south. This time, I reached only 500 feet when the engine quit. Again luck was with me. Enough snow was on the ground for me to land on the tundra with the skis. Again there was no damage to the plane. And again it required a complete engine change.

I wondered then if I might be wise to change my profession. After thinking it over I decided to continue flying. The three incidents, coming so close together, certainly got my attention. At the least, losing those engines kept me on my toes and emphasized the importance of defensive flying—that is, flying with my mind prepared for any emergency.

I also had new incentive to intensify my preflight inspections. I remembered an earlier incident, not in Alaska, that also made me sit up and take notice. A mechanic had just completed work on the brakes of a plane I was flying. I was taxiing out for takeoff when a wheel fell off. After that I always vigorously kicked the wheels on my preflight inspections.

IN MY FIRST YEARS OF FLYING from Bettles, CAA employees were mostly practical people who enforced flying regulations with common sense. Alaska's aviation industry was undergoing major growing pains at the time. Fine airports built during World War II by the military became available for civil use. The military had also built a Territory-wide communications system, which the CAA took over. And many large, modern military aircraft became available to Alaska's struggling airlines at bargain prices.

In the 1960s, by the time Alaska's airlines were on their feet and expanding, the CAA bureaucracy had grown in typical federal government fashion. Many CAA employees seemed to have little to do, and some seemed determined to impose major penalties for infractions that sometimes didn't exist.

For a time, the CAA's scrutiny was so severe that many pilots almost felt compelled to consult an attorney before every flight. Here is a case in point:

The Curtiss Commando, an Air Force plane also known as the C-46, was ideal for hauling freight in Alaska. It has a large cabin, ideal for bulky items, and its two 2,000-horsepower radial engines gives it tremendous performance. Many of these airplanes were available for lease or purchase.

During the early 1960s my close friend Bill Smith won a contract to deliver fuel oil to Bettles Field with his newly acquired C-46. I happened to be at my desk in the roadhouse when Smitty made his first landing with fuel at Bettles. As I watched, he set the big transport down in a gentle landing. However I couldn't resist the urge to pick up my radio microphone and kid him.

"Smitty, when are you going to learn to fly that airplane? You landed so hard you damaged our runway."

"Your eyesight gets worse every day, Andy," he joshed back. "One of these days they're going to ground you when you walk out of the roadhouse and can't find your airplane."

We had had our fun, and the radio talk was forgotten—or so we thought.

Several days later Sig Wien arrived at Bettles Field. I was flying a mail route, and when I returned Sig told me that during my absence a CAA safety inspector had visited Bettles, seeking a statement from me about Smitty's hard landing with the C-46.

At first that amused me. But when I considered the consequences, I had second thoughts. Someone had overheard Smitty and my radio joshing and had taken it upon themselves to report it to the feds.

When I told Sig the incident was all in jest, that I had nothing to report to the CAA, he suggested I call them anyway. I did so, only to be told that before Smitty purchased the C-46, another pilot had made a hard landing with the plane, resulting in minor damage to the landing gear.

Smitty hadn't noticed the damaged gear during his preflight inspection. The CAA imposed a healthy fine on Smitty for flying a damaged airplane. I felt badly about this, but I did learn never to make comments on the radio about anyone else's flying ability.

I HAD ANOTHER UNPLEASANT ENCOUNTER with the FAA (successor to the CAA) several years later. While on one of my flights to Anaktuvuk Pass, the crew of a Wien cargo plane off-loaded and scattered gasoline drums randomly in the Bettles airplane parking area. Unaware of this, on my return I taxied the four-place Cessna 185 I was flying into one of the drums and nicked the prop. Damage was minimal. I called the Wien operations officer and asked if he thought I should report the incident.

At the time the FAA listed two categories of damage to an aircraft. An "incident" was minor damage with minimal costs. FAA regulations did not require reporting an "incident." The second category was for serious damage, and reporting was mandatory.

The operations officer advised, "Andy, to be on the safe side report this as an incident, and it will probably be ignored."

Wien mechanics pulled the propeller, filed the nick out, ran a balance check, and put it back on the plane ready for service. Total cost, including labor, was $250.

I reported the incident in detail to the FAA. Thirty days later I received a call from the FAA legal department informing me that an FAA attorney had filed a civil suit against me, demanding a substantial fine.

I not only questioned this action, but asked for a jury trial. This created a standoff lasting most of a year. I wanted to settle the ridiculous dispute, and I went to the District Attorney who was attempting to collect the fine. He reviewed the case and got a look at the receipt showing the minor cost of repair to the propeller, and the case was dropped.

BUSH PLANES

A careless repair job on a bush plane once gave me a real fright. It came during a weeklong potlatch, or celebration, that the Koyukon Indians along the Koyukuk River usually hold at the end of beaver season, which is the end of trapping for winter. The potlatch draws people to the host village from other villages up and down the river.

Fur buyers show up to buy from trappers, and a lot of cash floats around. I often hauled planeloads of fur, mostly beaver, worth fifty thousand to one hundred thousand dollars to these potlatches. I also hauled passengers to and from the host village.

At potlatch time I usually asked Wien for use of the largest single-engine passenger airplane available. One year they sent me a Noorduyn Norseman, not the usual Norseman I kept at Bettles. This Norseman had four gasoline tanks. In order to keep the weight of the plane down, I left at least two, sometimes three, of the tanks empty.

One day while flying a full load of eight passengers to their home village, I was on a landing approach to a frozen lake. Suddenly the fuel pressure light went on and the engine sputtered. I had filled two of the fuel tanks and knew I had sufficient fuel, so I hastily

Richard Wien watches as I polish nicks out of the propeller of my favorite Cessna 180. We're at the Wien float on the Koyukuk River at Bettles, in 1955. Richard is the son of Noel Wien, one of the founders of Wien Airlines, and he also made Alaska aviation his career. JIM REARDEN PHOTO

switched the fuel control valve to what I thought was the proper position for the other tank. The engine continued to sputter.

At this point I doubted I could stretch my glide to reach the lake. I frantically switched the fuel tank selector to another position. Fortunately this resulted in full power, and I managed a safe landing.

I wiped the sweat off my brow, unloaded my passengers, checked the plane over the best I could, then flew back to the host village. I still had several hundred people to fly to their home villages, but I grounded the aircraft and called for a mechanic from Fairbanks to check the plane out.

That night, with flashlights, in 30-below temperature, the mechanic and I started to troubleshoot the plane. Since the problem centered around the fuel, we started by checking fuel screens. Ordinarily we would have drained the fuel tanks first, but we had no containers to drain the gas into, so I elected to remove the fuel screens as was.

While doing so, I spilled gasoline on my right hand and the flesh froze instantly. I ran for shelter and thawed the hand for an hour or so, and I believed there was no damage. About ten days later I lost all the fingernails on that hand. When the nails regrew, they were distorted and off color—a result of carelessness I have carried ever since.

We investigated the plane's maintenance log, searching for answers, and learned that a previous pilot had reported that the plaque that designates which fuel tank is in use had loosened. The mechanic who repaired it had failed to line it up with the proper markers. This made it impossible to determine which of the four tanks was feeding fuel to the engine.

My hair turned gray early in life, and I think this incident added a few gray hairs. We repositioned the plaque, and I had no further trouble with that plane.

AFTER A FEW YEARS OF FLYING in Alaska's Arctic, I held strong views on the traits that make for a good bush plane. Ruggedness is important, because a bush plane must often land and take off on uneven ground, snow, or rough water. Heavy loads and a wide diversity of loads from game meat to mining equipment are common, and a lightly built plane doesn't stand up under such use.

I prefer a single engine—the controls are easier to handle in dicey takeoffs and landings. And a good bush plane must be able to handle short takeoffs and short landings.

A really efficient bush plane should have a door and cabin large enough for loading a 4-by-8 sheet of plywood and standing it on edge. When I first started flying the bush, my planes could carry sleeping bags or folding cots but not regular beds. In later years I could deliver king-size springs, mattresses, and beds.

One would think that two apparently identical planes would have exactly the same flying characteristics. Perhaps some do when they leave the factory, but it seems to me that each airplane develops its own traits—almost personalities. I had a favorite, a Cessna 180, a four-place plane powered by a 225-horsepower Continental engine. Wien sent this plane, N9311C, to me at Bettles when it was brand new. I didn't realize what I had until it was fitted with floats for the summer.

Floatplanes especially seem to develop their own idiosyncrasies. Over the years I had to experiment with each floatplane I flew to learn how to obtain the best performance. Most airplanes are sluggish when flown with floats.

Cessna N9311C was a pleasant surprise. The more I flew it, the more confidence I gained in it. Normally it is necessary to drastically reduce the payload of a plane, especially one on floats, in order to

One of my favorite bush planes was this responsive, highly dependable Cessna 180. We're on the shore of Chandalar Lake in the Brooks Range in this photo from September 1955. JIM REARDEN PHOTO

gain performance. Not so with One One Charlie. I flew the plane for two years for a total of 4,000 hours and in all that time on only one occasion did I have to reduce my load to take off.

I could fly this airplane from a surprisingly small lake—so small that I often drifted into a bank when landing in order to stop. The floats installed on the plane were the Edo 2425 model, and with a gross load they would be almost submerged. Yet when I applied power, the plane quickly lifted onto the step and flew off the water beautifully.

On only one occasion did the plane refuse to get off with a gross load. It was at ten-mile-long Chandalar Lake at 3,222 feet elevation on a miserable day with almost no ceiling and poor visibility. I was on a regular mail run, and it was my last stop. A postal inspector was with me, making one of his regular rounds. I picked up two additional passengers—Cappy Adney, who was going on vacation with umpteen pieces of luggage, and a hunter who had bagged a Dall sheep and a caribou and was returning to Fairbanks.

I loaded the three passengers and their luggage, mail, personal gear, plus the meat and horns and antlers of one sheep and one caribou. It was windy with big swells on the lake. We were so low in the water I had difficulty keeping the nose up to clear the prop from the swells. After several futile attempts to get on the step I was forced to return to the beach to reduce the load.

But I had a problem; there was no beach I could taxi up on. The lake has an abrupt shoreline with deep water a few feet out. I feared the tail of the aircraft might sink once the floats hit the bank and I cut the engine to off-load passengers .

To be on the safe side I eased onto the bank, left the engine running, and told the passengers to step onto the floats and move as far forward as possible, but to be sure to stay behind the wing lift strut and away from the spinning prop.

They followed my directions perfectly. I then tied a line to the rear of the float, cut the engine, jumped ashore and drew the tail of the plane onto the beach. No harm was done, except for my embarrassment.

I off-loaded about a hundred pounds, reboarded everyone, taxied out into the lake, applied power, and old One One Charlie responded like the champion she was and we were off and flying back to Bettles. I was a little wiser after that experience.

WHEN I FIRST STARTED TO FLY for Wien, the company was short of airplanes, and I flew whatever was available. I had problems with the Republic RC-3 Seabee that Wien came up with. The next plane was a Piper Family Cruiser. After that came the Cessna series, starting with the two-place, 90-horsepower Cessna 140. Then came a four-place Cessna 170 that used a 145-horsepower Continental engine. I remember once when Wien owned eight of these airplanes and six were grounded with engine problems.

The next step up was to the Cessna 180 with a 225-horsepower Continental engine, which proved to be a reliable workhorse I flew for many years. There were times, though, when I flew a Twin Beech, Twin Cessnas, a Stinson Jr., the Noorduyn Norseman, and later several DeHavilland Beaver and Otter aircraft. The Beaver, designed and built in Canada for bush flying, is a very fine airplane. Though no longer built, many still fly in Alaska and elsewhere.

All of the Wien pilots as well as the top administrators of the company were constantly on the lookout for the ideal bush airplane. In the early 1960s I happened to read a story about a mountain climbing expedition in India supported by a helicopter and a Pilatus Porter fixed-wing airplane. The account detailed how the pilot of the

Porter, flying at 19,000 feet, was stricken with hypoxia (lack of oxygen). He had to land, and the only available place was an area with an accumulation of boulders as large as the plane. There he had stalled the plane into the rock pile. A photo of the landed plane showed little damage. I was struck by the size and performance of the Porter.

I learned from the article that the plane was built in Bern, Switzerland. It was powered by a constant-speed French Astazou turbo engine that operated at 90 percent power and rotated a thirteen-foot-diameter electrically controlled, variable-pitch propeller. The propeller was reversible, making it possible to land and stop on a dime. Further, it carried eight passengers plus pilot and had large double doors on each side of the fuselage.

Was this the ideal bush plane? I was so impressed that I sent the article to Sig Wien. Soon he was on his way to Switzerland to visit the Pilatus Porter factory. He ordered a Porter airplane with a conventional 300-horsepower, Lycoming piston aircraft engine. The plane was disassembled in Switzerland and flown by air transport to Long Island, New York, where it was reassembled and flown to Alaska, arriving in 1962.

The airplane had a remarkable load capability, but I was disappointed by its performance with the Lycoming engine. Sig Wien soon placed a second order for a Porter, this one to be equipped with the French Astazou turbine.

Because the turbine was so much lighter than the conventional piston Lycoming engine, an additional fifty-one inches of length had to be added to the nose of the aircraft to keep it in balance. My first flight with the turbine-powered Porter left me gasping; its astounding performance opened for me a whole new world of heavier and bulkier loads and short field operation.

But there was a drawback. Engine starts were fully automatic. If

The Pilatus Porter was a great bush plane, built in Switzerland, that I flew during the early 1960s. It carried eight passengers and had large double doors on each side of the fuselage.

the battery wasn't fully charged, the engine failed to ignite. That was bad enough, but the failure resulted in jet fuel being delivered to the engine burner. If the engine wasn't successfully purged of this unburned fuel, at the next start it would be loaded with a double dose of fuel. When this was ignited, a ten- to fifteen-foot sheet of flame shot out of the exhaust. In seconds, the double whammy of jet fuel pushed the engine exhaust temperature above 1,000 degrees Fahrenheit.

If exhaust temperature exceeded 1,000 degrees, it was mandatory for the engine to be removed from the plane and sent to the factory for inspection. Since a replacement engine cost forty thousand dollars, hot starts had a tendency to increase operating costs. And how.

I occasionally telephoned the Astazou factory in France for

information, for parts, whatever. For parts or ordinary information, we had no problem communicating. But with engine trouble calls, suddenly no one at the factory could speak English, nor could anyone understand our problem.

Our salvation, which permitted us to continue using the Porter, came from a good old American competitive product. Pratt & Whitney developed a turbine engine that eliminated the majority of engine problems we were experiencing. Plus, the Pratt & Whitney PT6A-6 gave even better performance than the French engine when installed in the Porter.

With the wonderful performance and reliability of the Pratt & Whitney-powered Porter, Wien was able to retire some of its smaller bush-type airplanes while taking on additional work. Wien eventually owned eight of these grand airplanes, four of them with the Pratt & Whitney turbo engine. The planes cost about $100,000 each.

I rapidly built time in the Porter, and on one occasion the Pilatus Corporation recognized me as the highest-time Porter pilot in the world. It is an absolutely marvelous airplane that can fly almost unbelievably heavy loads.

The heaviest load I ever hauled in a Porter was a huge cast-iron water pump for Kennicott Copper, which was conducting exploratory drilling in the Kobuk area. The pump was delivered to Hughes by a large transport plane with a request that I pick it up and fly it to Kennicott's Bornite camp.

The perfect windless day arrived, and at Hughes I asked Abraham, the Wien attendant, to load the pump into the Porter with his hydraulic tractor lift, setting it as far forward in the cabin as he could manage. I asked him to tie it down.

He reported back that he had loaded the pump, but couldn't get the airplane doors shut. I discovered I had misjudged the weight of the pump; it was so heavy it had bowed the fuselage until the doors wouldn't close. I had Abraham close the doors as far as they would go and tie them shut with rope.

I was apprehensive when I taxied the Porter to the end of the 4,000-foot runway at Hughes. Would it fly with that terrible load? I needn't have worried, for the airplane handled the weight beautifully. I had a smooth flight, with nary a bump.

Next I had to land. I knew I had to make a power approach and touch down as gently as possible. But the field at the Bornite camp was only 1,200 feet long with a dogleg in the center. I decided to land instead at the nearby airfield built by the Dall Creek Mining Company. I landed gently, and with considerable relief parked at the end of the runway.

When the pump was unloaded, the airplane fuselage sprang back to normal and the doors closed properly. I flew back to Bettles with a light heart—and a much lighter airplane.

WIEN OFTEN SENT NEWLY HIRED PILOTS to Bettles to get experience and to accumulate flying time, especially those who were expected to fly the bush. Such was the case when the company sent an ex-Air Force pilot to Bettles to be checked out in the Porter and to build some bush flying time.

On a beautiful Sunday morning I decided to fly a load of freight to Bear Creek Mining Company, and I thought it a good time to give the new man some Porter time. He did a good job flying the plane from the right seat—so good that I asked if he'd like to make a landing at Bear Creek. He assured me he would, and we changed seats, putting him in the left-hand seat, me in the right. Brakes for the Porter are mounted only on the left.

We reviewed the landing checklist, including use of flaps and locking the tail wheel. When we reached Bear Creek, the new pilot's letdown and approach to the field were letter-perfect, as was his touchdown.

The Porter has a tendency to veer left on landing. Usually a little right brake corrects this. After touchdown, when the plane started to veer left, I called out, "Right brake. Give it right brake."

Suddenly the plane veered sharply left, slamming against a rock that was about as big as the airplane. In about five seconds, one hundred thousand dollars flew out the window, for the airplane was mortally wounded.

I asked the pilot what had happened, as everything had gone smoothly until the plane veered left. The following day I questioned him again and he mentioned something about the tail wheel. I asked him to repeat what he had just said. "Since the plane wanted to veer left, and you kept yelling 'right brake, right brake,' I misunderstood, and unlocked the tail wheel."

It was now obvious what had caused the plane to veer left so violently. Once the tail wheel is unlocked it sets up a sling effect; with a slight veer, the tail follows quickly unless the tail wheel is locked. A heavy load accents the action.

The Porter was disassembled, loaded into one of our cargo planes, and hauled back to Fairbanks. It never flew again.

FAREWELL

In 1956 I went to Europe with Sig Wien and the airline's operations manager, Dick King, to look for new aircraft. We wanted a plane to replace the Douglas DC-3s that handled the traffic for our main routes.

We learned of the Fokker F-27, built in Amsterdam. We looked these planes over and talked to the Fokker people. Eventually Wien purchased a number of these fine planes. We bought a couple of twin-engine bush-type planes to add to our ever-growing fleet.

With new and more modern planes, the door opened for a booming tourist business. Wien established tourist hotels at Nome, Kotzebue, and Barrow. By this time I was a senior pilot and I could have bid for a position on the main line flying out of Fairbanks, but I was enjoying my life at Bettles and had no desire for a change. I gained much satisfaction in providing service to the villages and mines, where I had many close friends. And business continued to build in the Koyukuk Valley.

For ten straight years I flew more hours annually than any other Wien pilot. In the seventeen years I flew from Bettles Field, I accumulated 32,000 hours of logged flying time. The most hours

For a period in the 1960s, I put on coat and tie and worked as executive vice president of Wien Airlines, in Fairbanks. WIEN AIRLINES PHOTO

I flew in any month was 276. At one time I flew 600 hours in a three-month period.

In the early 1960s, though business seemed to be booming, Wien Airlines wasn't doing well financially. By this time I had a considerable investment in the airline, having accepted stock in lieu of cash for many services and expenses. Also over the years I had advanced money to the company for investments in aircraft and other equipment, accepting stock in payment.

I flew to Fairbanks and talked with Sig Wien, who frankly told me he was deeply worried. He even feared losing the airline. We talked for several hours, and finally Sig suggested, "Andy, why don't you come to Fairbanks and take over operation of the airline."

This was an unexpected bombshell. Not only did I not want to be an executive running a big operation, but it also wasn't the plan I had for my life. "Sig, I'm only a few years from leaving Alaska to

246

Sigurd Wien, pictured here in the mid-1960s, was one of the founders of Wien Airlines and served as chairman of the board of directors for more than thirty years. He died in 1994 at the age of ninety-one. WIEN AIRLINES PHOTO

return to my Pennsylvania farm," I told him. "My deadline, which I set when I arrived here in 1947, is 1967."

After much persuasion, I agreed to leave Bettles and move to Fairbanks to become executive vice president and work with Sig. We would be a joint management team. We would share the same office, look each other in the eye each morning, and together decide on each day's activity for the airline.

It was a traumatic time. Changing the old ways wasn't easy, and change upset some longtime employees. I probably made some mistakes. I know there were hard feelings from some. Many times I regretted the decision to work with Sig at Fairbanks and longed to be back flying in the Koyukuk country.

Within two years, however, the company had made a substantial profit.

In 1965 Wien Airlines and Northern Consolidated Airlines merged. Northern Consolidated, like Wien, had provided service for many years to bush communities in Interior and Southwestern Alaska. At the merger the new company was admitted to the New York Stock Exchange.

In order to make the stock more attractive and affordable, the common stock was split 52 to 1. I suddenly found I was the third-largest stockholder of the combined companies, now known as Wien Consolidated Airlines, with about 1,400 employees and more than fifty airplanes.

From the time of my arrival in Alaska in 1947 I had made clear to everyone, and promised myself, that I would put twenty years into my Alaska career, then return to my Pennsylvania farm.

When my parents bought the farm, the old home had gas lights, individual gas heaters in each room, and no water or indoor plumbing. In 1956 I hired a contractor to restore the house and to add electricity, plumbing, and central heating. After that my parents lived there in relative comfort.

My father passed away in 1962, but my mother still lived at the farm. As my twenty years in Alaska neared an end, she was becoming frail and needed assistance. I wanted to be there to give her that assistance.

I simply could not convince anyone that I was serious. Nor could I convince Sig and others at Wien Airlines to hire anyone to take my place. Nevertheless I gave my notice.

ON THE FIRST DAY OF JUNE, 1967, I didn't show up at Wien headquarters for work; instead I was headed south, driving down the Alaska Highway in a Chevrolet truck loaded with my personal possessions.

My farm-life dreams were not shared by my wife, Hannah. She decided to remain in Alaska, and I moved to Pennsylvania. This was the most difficult of my life's decisions. With children involved, both Hannah and I were torn in many directions. The children remained with Hannah, but my focus included financial care of all of them until the children completed school and were trained to be independent.

To this day I feel guilty over this, because material provisions can never replace the personal touch when raising children.

My life in Pennsylvania was one of relaxation, a major change after nearly two decades of constant pressure to meet airline schedules and making decisions affecting other people's lives. I was only forty-five years old, and for the first few months I pondered over what I would do with the rest of my life.

Farming as I practice it is mostly a summer activity, leaving much of the year without much to do. I tried starting a small flying service at the local airport, but found it so different from Alaska flying that I quickly lost interest. In Alaska, flying is a lifeline; in Pennsylvania it is mostly a luxury since so many competing transportation services are available.

My mother passed away in 1968, one year after my return to the farm. I was thankful for my decision to leave Alaska, giving me a few months with her during her last days.

That fall I received a call from the executive officer of a firm in New York that had invested heavily in Wien Consolidated Airlines. Would I return to Fairbanks to help manage the company? I would work directly with them to protect their investment.

I did return to look things over, since I was still a major stockholder. I decided that I wanted no part of it and resigned my position on the board of directors.

Wien Airlines (the "Consolidated" was dropped) attempted to expand jet passenger service into the western states, competing with bigger, long-established airlines. Subsequently a pilot's strike and poor management (not by any of the Wien family) resulted in failure of the company and it went out of business.

The words "integrity" and "Wien Airlines," at least into the early 1980s, were synonymous. The company's growth was slow and at times painful. In the early years, employees at times were asked to defer cashing salary checks for as long as four months because there simply wasn't enough money in the bank to cover them. But they were always made good, and bills were always paid; a promise made by Sigurd or Noel Wien was as good as a written contract.

Those who criticized the company probably weren't aware of the huge unpaid accounts due the company. This was due to the policy of providing service to any and all, regardless of ability to pay. An unwritten goal of Wien Airlines while I was with the company was to improve the living standards of their rural customers—especially the Indian and Eskimo villagers. This was accomplished by building aircraft runways, establishing a village radio network, and providing dependable flying service. It was a sad day for me and for many Alaskans when Wien Airlines went out of business.

Sig Wien, who for decades was the backbone of the company, was a loner, with few intimate friends, and I felt privileged to be close to this pioneer Alaskan aviator-businessman. He was a student of the Bible, and his particular faith didn't permit him to recognize Christmas. Yet he often arrived at Bettles for holidays, including Christmas. And there had better be a few gifts under the tree for him.

When Sig Wien died in December 1994 at the age of ninety-one, I lost one of my best friends. He was a quiet, efficient man, who helped make the Wien name the most famous in Alaska aviation.

MY PENNSYLVANIA FARM is many thousands of miles from Alaska, but part of me still resides in that wonderful land. My daughter Mary lives in Fairbanks with her two daughters: Kelly, a champion swimmer, and younger Lacey. My oldest son, Phil, also lives in Fairbanks, and is a teamster driving a heavy truck rig on the North Slope. He is also active with Alaska Native Corporation business.

My other son, David, became a professional baseball player and spent two years pitching for Dodgers farm clubs. He injured his arm and was forced to retire from playing. He now lives in Las Vegas and works in the freight department for American Airlines. As a sideline, he coaches high school baseball. He has one son, Casey, who lives with his mother in Reno.

I remarried, and I share a happy life on the Pennsylvania farm with Betty, my wife now for more than thirty years. An ailment slowed me down for a while, but I'm doing fine for now. The ailment showed up a couple of years ago after I had a minor accident while cutting wood on the farm. My head started bleeding, and a medivac helicopter flew me to a Pittsburgh hospital (how many times did I fly injured or ill passengers to a hospital?). Six weeks passed for me at the hospital, none of which I remember. My doctors told me my heart stopped three times during that period, but each time they brought me back to life.

I don't know what the good Lord has planned for me, but if the next seventy years are as exciting as the first seventy, I'll be a happy man. The years since I left Alaska have been good ones. Nevertheless, when I see a small plane high in the sky, or when I think of that magnificent land far to the north, contemplate its clear rivers, remember skyscraping, snowcovered peaks, abundant wildlife, and the wonderful people I once knew, I feel a longing, a homesickness, for the magic years when I flew as an arctic bush pilot.

ABOUT JAMES L. ANDERSON

James L. Anderson, born in 1922, grew up in the coal-mining town of Montcoal, West Virginia. He became an Eagle Scout, played on the high school football team, did odd jobs, and worked in the Montcoal electrical power plant.

He enlisted in the Navy five days after Pearl Harbor, and in six years of service advanced from apprentice seaman to lieutenant (jg). After flight training, he was assigned to the aircraft carrier USS *Princeton,* from which he flew Curtiss dive bombers in combat. In 1946 he returned to civilian life and for a year became a Civil Aeronautics Administration flight controller at Bettles in the Koyukuk Valley of northern Alaska.

Realizing the need for bush plane service in the Koyukuk, he resigned from the CAA and, backed by pioneering Wien Airlines, almost single-handedly established regular, dependable air service for villages, mines, and travelers in a vast region of northern Alaska. He built a roadhouse, or lodge, at Bettles, which became the focal point of commerce in the Koyukuk region.

In the seventeen years he flew from Bettles Field, Anderson accumulated 32,000 hours of logged flight time in a wide variety of single- and twin-engine airplanes. After twenty years of aviation in Alaska, he retired to a Pennsylvania farm, where he lives with his wife, Betty.

ABOUT JIM REARDEN

A fifty-year resident of Alaska, Jim Rearden has written seventeen books and more than five hundred magazine articles, mostly about Alaska.

Recent books include *Shadows on the Koyukuk,* the life story of trapper, businessman, and public-service leader Sidney Huntington; *In the Shadow of Eagles,* on the life of barnstormer/bush pilot Rudy Billberg; *Koga's Zero,* the story of the first Japanese Zero airplane captured and flown by the United States during World War II; *Alaska's Wolf Man,* relating the wilderness adventures of Frank Glaser (for which Rearden was named historian of the year in 1999 by the Alaska Historical Society); and a novel, *Castner's Cutthroats,* about the famed Alaska Scouts of World War II.

Rearden earned a BS degree in fish and game management from Oregon State College and an MS in wildlife conservation from the University of Maine. In Alaska he has served as a federal fishery patrol agent, taught wildlife management at the University of Alaska Fairbanks, and was fisheries biologist in charge of commercial fisheries in Cook Inlet for ten years. He has also been a commercial fisherman and a registered big-game guide. He served on the Alaska Board of Game for twelve years. He is a private pilot.

Rearden was outdoors editor for *Alaska* magazine for twenty years, and for twenty years he was also a field editor for *Outdoor Life* magazine. He lives in Homer with his wife, Audrey, in a log house he built.

Other books about bush pilots
of Alaska from Epicenter Press

FLYING COLD: The Adventures of Russel Merrill,
Pioneer Alaskan Aviator, Robert Merrill MacLean &
Sean Rossiter, softbound, $24.95, hardbound $34.95.

ALASKA'S SKY FOLLIES: The Funny Side of Flying
in the Far North, Joe Rychetnick, softbound, $13.95

To order single copies, mail the purchase price (WA residents
add $1.46 sales tax) plus $5 for Priority Mail shipping to:
Epicenter Press, Box 82368, Kenmore, WA 98028; phone our
24-hour order line, 800-950-6663; or visit our website,
EpicenterPress.com. Visa, MC accepted.